An Introduction to Abstract Algebra

Alan Parks
Lawrence University
Appleton, Wisconsin

♠ This project puts text material, carefully coordinated with lectures, homework, and bibliography, into the hands of upper level students at cost. It was begun during the 1996-97 academic year, and it has been revised and expanded periodically since.

©Alan Parks. Anyone can copy or excerpt this work, provided that this page and the previous cover page are included in their entirety and provided that no cost is associated with the copy or excerpt in any form on any medium. The use of such copy or excerpt must include a reference to this work as

Parks, Alan, *An Introduction to Abstract Algebra*, 2nd Ed., $\alpha\lambda\alpha\sigma$ Publishing, Appleton, Wisconsin, 2018.

This license transfers to any such copies or excerpts. Other than the specific actions allowed in this License, the copyright holder retains all rights. The principal copy of this edition was printed on December 27, 2018.

Contents

Introduction		v
Chapter 1.	Logic and Sets	1
Chapter 2.	The Integers	11
Chapter 3.	Modular Arithmetic	23
Chapter 4.	Functions	31
Chapter 5.	Permutations	45
Chapter 6.	Groups	57
Chapter 7.	A Single Group Element	67
Chapter 8.	Factorization	77
Chapter 9.	Subgroups	89
Chapter 10.	Lagrange's Theorem	97
Chapter 11.	Normal Subgroups	105
Chapter 12.	Group Homomorphisms and Isomorphisms	119
Chapter 13.	Conjugacy Classes	135
Chapter 14.	Commutative Rings	145
Chapter 15.	Residue Units	155
Chapter 16.	The Rational Numbers.	165
Index		175

Introduction

"Those who undertake to write take that trouble for many reasons; for some of them apply themselves to show their skill in composition; but others, of necessity and by force, are driven to write because they are so concerned with the facts, that they cannot excuse themselves committing them to writing."

Josephus.

It is easy to recognize mathematics. There is an explicit and complete list of assumptions, the definitions are formal and unambiguous, calculations and deductions follow the rules of logic unfailingly, the conclusions are unavoidable and unassailable. Now, a given argument may make *some* of its assumptions clear, may use *certain* of its terms carefully, may present conclusions which are *more or less* convincing. In this sense one argument may be *more mathematical* than another, but there is really no middle ground at all between arguments that are formal mathematics and those that are not. It is a target with only the bull's-eye and nothing else. Once the ancient Greek geometers saw how many interesting consequences could be derived rigorously from a few rather innocuous axioms, the idea spread that the constraints of mathematical method could be applied to an endless variety of problems. And they were and are.

The insistence on correctness and formality at every level of a mathematical argument might lead one to believe that the subject is cold, mechanical. Certainly the proof of some particular fact, or the solution to some particular problem, should have a certain coolness. But it is the cool-headedness of certainty, not that of personal indifference. Proofs themselves, the finished products of mathematical investigation, come about only as a very human mind (yours!) grapples with the issues at hand. An electrical wire is cold in the sense that it conducts electricity rather effortlessly, but if the point is to light up the room then we want to waste as little energy as possible in conducting the current. A mathematician wants to waste as little doubt as possible (none!) on the certainty that mathematical results follow from their premises.

This text introduces abstract algebra for a course in which you will begin learning to read, write, remember, and do mathematics. To this end we will study elementary number theory and the theory of groups – subjects fundamental not just to advanced mathematics, but to virtually every discipline of quantitative science. More importantly from our point of view, they are intrinsically interesting, and we can prove results of significance starting from scratch and working in a self-contained way ideal for providing firsthand experience with mathematics.

Firsthand experience will occur only as you diligently pursue mastery of the course content in order to gain problem solving ability and the skill of algebraic work. You will not benefit from the text unless you read it very aggressively, using your ability to reproduce the various arguments to test understanding. To repeat: you do not *know* or *understand* a subject in mathematics unless you can work out its main theorems from the ground up on your own. Thus, mathematical exposition tends toward the sparing side. It is assumed that the reader works through the arguments thoroughly and carefully.

In class we will work hard at understanding the theorems proved in the text, and we will work a variety of problems, showing how to apply the material.

At the end of each chapter, you will find select bibliography for many of the text topics. A library search over classifications QA150-272 will yield many, many texts that cover the material in this book, so you can find alternative points of view, additional problems, and other topics, if you wish.

CHAPTER 1

Logic and Sets

Mathematics is easy to describe: you have a list of statements that you accept as true, and you construct examples and derive logical conclusions from those statements. The original statements are called *axioms* and the conclusions are called *propositions* or *theorems*. That's all there is to it! The subject seems to have arisen in Greece around 600 BC, and wherever it has been introduced it has absorbed whatever tentative progress had been made previously in its direction.[1] Learning advanced mathematics consists in becoming aquainted with axoims and theorems that are simultaneously useful, interesting, and beautiful, as judged over many centuries by people all over the world.

Mathematicians think in at least two ways; we might say *formally* and *intuitively*. Every mathematical argument is a formal display of precisely defined objects and strictly logical manipulations. Without precision and clarity you do not have mathematics, and to learn mathematics you need to learn to recognize when an object is well-defined, when a proof is correct and complete. But the objects and logical bindings are not interesting and not memorable unless they correspond, intuitively, with notions in the mind that are familiar and appreciated. Furthermore, you will never make much progress in producing mathematics unless you learn to think intuitively about the objects and arguments, creatively employing various metaphors or pictures. Visual or

---

[1] One of many books on the history of mathematics is *A History of Mathematics* by Carl Boyer, Wiley, 1991.

poetic images bring abstract objects to life. The skill of the mathematician consists in being able to combine these approaches: to be able to imagine a formal, abstract object pictorially; to be able to translate intuitive thought into formal language.

On the formal level, the positive integers form a set with certain definite properties (we will write down these properties momentarily). Intuitively, you have all sorts of thoughts in your mind involving the positive integers: you might remember the manipulatives you used in learning to count, you might imagine marks on a number line, you might visualize $2 \cdot 3 = 6$ as counting the entries in a table with 2 rows and 3 columns, etc. Formally speaking, understanding a proof about the positive integers is an exercise in matching clearly defined properties with logical statements about them; intuitively speaking, understanding a proof is an exercise in seeing a pattern of sensible thought. The important facts are not arbitrarily chosen, rather, they are aesthetically interesting, displaying a sense of beauty or elegance, whether we are talking about the practical beauty of the solution to an applied problem or the ethereal beauty of a symmetry in music or a painting. Aesthetic appreciation and formal understanding go hand in hand.

To prepare for the integer axioms, we introduce some logic. Much of this may be familiar to you; we need to make sure we understand the use of logical terms as they are used in mathematics.

Mathematical statements[2] are put together with logical connectives that we now list. If we are given a statement **A**, then its *negation* is denoted "not **A**." The negation of **A** is false if **A** is true, and it is true if **A** is false.

---

[2] We will approach the idea of a statement via examples. For now it suffices to say that all statements considered in this course will be either *true* or *false*, and not both. It turns out to be possible for a statement to be neither true nor false – and not because it is ambiguous. The general issue lives in the realm of *mathematical logic*.

If we have statements **A** and **B**, then the statement "**A** *and* **B**" is true if both **A** and **B** are true, and it is false if either **A** or **B** is false.

The statement "**A** or **B**" is true if either **A** or **B**, or both, is true, and it is false if both **A** and **B** are false. It is worth noting that this use of the word "or" is somewhat different than that of common English. If, in everyday speech, I say, "I am going to the store, or I am going to a movie," I probably do not mean that I am both going to the store *and* that I am also going to a movie. If I did mean that both were true, I would have said, "I am going to the store and to a movie." In mathematics, however, the word "or" allows either or both its constituent statements to be true.

A statement of the form "if **A**, then **B**" is an *implication*. This statement can be written more symbolically: $\mathbf{A} \Rightarrow \mathbf{B}$. In such a compound statement, the statement **A** is the *hypothesis* and **B** is the *conclusion*. An implication is true if both its hypothesis and conclusion are true, and it is also true if the hypothesis is false, regardless whether the conclusion is true or false. Thus, the statement "if $2 + 2 = 4$, then Paris is a city" is true, as is the statement "if $1 = 0$, then pigs have wings." An implication like the second one, in which the hypothesis is false, is said to be *vacuous*. There is an important role for vacuous implications, as we will see later.

The logical value of statements involving *and, or, implies* can be represented by *truth tables*, where T stands for *true* and F for *false*. The third line after the heading of the table says that if **A** is false and **B** is true, then (**A** and **B**) is false, (**A** or **B**) is true, etc. The idea is that the T's and F's in columns 3, 4, and 5 follow from the T's and F's in columns 1 and 2.

A	**B**	(**A** and **B**)	(**A** or **B**)	(**A** $\Rightarrow$ **B**)
T	T	T	T	T
T	F	F	T	F
F	T	F	T	T
F	F	F	F	T

You can test your understanding of what we have done so far by showing that the statement "**A** $\Rightarrow$ **B**" is *equivalent* to "(not **A**) or **B**." (Equivalent statements are both true or both false.) Simply make a truth table column for ((not **A**) or **B**).

As we will point out in class, we need to distinguish between *proving* that an implication is true and *using* an implication that is already known to be true.

We now consider the *contrapositive* of an implication. The contrapositive of the implication **A** $\Rightarrow$ **B** is the implication "(not **B**) $\Rightarrow$ (not **A**)." We aim to show that an implication is equivalent to its contrapositive (they are both true or both false). As above, this is simply a matter of truth tables. Here we go.

A	**B**	**A**$\Rightarrow$**B**	not **B**	not **A**	(not **B**)$\Rightarrow$(not **A**)
T	T	T	F	F	T
T	F	F	T	F	F
F	T	T	F	T	T
F	F	T	T	T	T

We see that the value of **A**$\Rightarrow$**B** is always the same as that of

$$(\text{not } \mathbf{B}) \Rightarrow (\text{not } \mathbf{A})$$

and that is precisely that the statments are equivalent.

The *converse* of the implication **A** $\Rightarrow$ **B** is the implication **B** $\Rightarrow$ **A**. In general, an implication can be true without its converse being true, and an implication can be false without its converse being false. (Can you think of an example?)

Notice that we can express that **A** and **B** are equivalent by "**A** $\Rightarrow$ **B** and **B** $\Rightarrow$ **A**." (Truth table!) We often say that **A** and **B** are equivalent by saying "**A** if and only if **B**." This is also written **A** $\iff$ **B**.

Next we turn to what are called *quantifiers*. Consider the statement "Every cow has four legs." The phrase "every cow" is a universal quantifier; let's see what this means by giving several mathematical ways to understand the statement. First, a formulation in logic:

If C is a cow, then C has four legs.

The idea of "every cow" is embodied in the hypothesis "if C is a cow." Next, a formulation in terms of sets – in terms of collections of objects. Let Q be the collection of all cows. The statement $C \in Q$ says that C *is an element of the set Q* – in other words, C is a cow! Let F be the set of all four legged animals. The sentence "every cow has four legs" looks like this.

If $C \in Q$, then $C \in F$

Of course, this is very similar to the logical formulation – both are expressed as implications. Finally, a more abstract version that introduces the formal symbol C before we know what it stands for:

$$\forall C(C \in Q \Rightarrow C \in F)$$

The symbol $\forall$ is read *for all*; it is a synonym for *every*. The introduction of C via the universal quantifier is like the declaration of a variable in a computer language. In practice, we will usually avoid the for-all symbol in favor of English words that indicate the logic.

What is the negation of "every cow has four legs"? We can reason it out fairly easily: we need to produce a particular cow that does not have four legs. It is instructive to see this in terms of the making false the implication "$C \in Q \Rightarrow C \in F$." In order for the implication to be false, we need the hypothesis to be true and the conclusion false. In other words, we need $C \in Q$ to be true (we need to be given a cow) but with $C \notin F$ (C is not in the set F; we mean that C does not have four legs). Notice that C is not univeral any longer; it doesn't refer to all cows; we are talking about the existence of a particular cow without four legs:

there exists a cow without four legs

In symbols:
$$\exists C(C \in Q \text{ and } C \notin F)$$

The symbol $\exists$ is read "there exists." That quantifier is called the *existential quantifier*. As with the universal quantifier, we far prefer English words (there is a cow...) to the symbol $\exists$.

Here is an abstract version of what we just did. Let **A** be a statement that involves the variable X. Here are the quantifier statements and their negations.

statement	negation
$\forall X(\mathbf{A})$	$\exists X(\text{ not } \mathbf{A})$
$\exists X(\mathbf{A})$	$\forall X(\text{ not } \mathbf{A})$

Here are the same statements the way they will occur in our writing:

Universal Statement	For every X the statement **A** is true.
Negation	There is an X for which **A** is false.
Existential Statement	There is an X for which **A** is true.
Negation	For all X, the statement **A** is false.

It is customary to give exercises that compound logical connectives and quantifiers. We will give a couple examples in class, and you will see such problems in the chapter exercises.

We have mentioned *sets*. A set is a collection of objects; if A is a set and B is any object we write $B \in A$ when B is one of the objects in the collection called A. We write $B \notin A$ when B does not belong to A's collection.[3]

Because we have mentioned *sets*, we take the opportunity to give three definitions related to them. The *empty set* ϕ is the set with no elements. It comes up among sets the way zero comes up in arithmetic, and so we want to have a symbol for it. We can use the universal quantifier to say that ϕ is empty: $\forall x(x \notin \phi)$.

---

[3]If you are starting to think mathematically, it will occur to you to ask, "What is an *object*?" We will not go into this very far, except to say that most mathematicians start with the idea of a set as axiomatic, and they say that all mathematical objects are sets.

Second, if the elements of the set A are also elements of the set B, we say that A is a *subset* of B. In symbols: $A \subseteq B$. That A is a subset of B can be expressed as the implication $x \in A \Rightarrow x \in B$. The empty set is a subset of every set – this is a use of the vacuous implication. Every set is a subset of itself, as well.

Third, if A and B are sets, we denote by $A \cap B$ the *intersection* of A and B. This is the set of elements in both sets:

$$x \in A \cap B \iff (x \in A \text{ and } x \in B)$$

Now that we have some logic and language, let's give our first proof. We'll start with a very simple fact that will probably seem obvious. We have annotated the proof extensively to present general principles that will apply to all our proofs.

PROPOSITION 1.1. *Let A, B, C be sets. Let $A \subseteq B$ and $B \subseteq C$. Then $A \subseteq C$.*

PROOF. The statement "$A \subseteq B$ and $B \subseteq C$" is our hypothesis. The conclusion is "$A \subseteq C$". The word "Let" tells us to regard the hypothesis as true. "Then" tells us that we should prove the conclusion.

The conclusion $A \subseteq C$ is itself an implication: if $x \in A$ then $x \in C$. Notice that we now have two conclusions on the table.[4] Here is an important principle: to prove an implication, *we assume that its hypothesis is true*. Why do we do that? Because if the hypothesis is false, the implication is true vacuously, and there is nothing more to do in that case. Thus, we assume that $x \in A$. The symbol x now represents a specific (but unknown!) element of A. Technically, it has been brought into our world by an existential quantifier.

---

[4]The entire statement $A \subseteq C$ is the conclusion of the Proposition we are proving. The conclusion $x \in C$ is the conclusion of that first conclusion!

We are assuming that $A \subseteq B$. That, too, is an implication: if $y \in A$, then $y \in B$. In this implication, the y is *not* a specific element of A, but merely a placeholder telling us that if $y \in A$, then the same y is an element of B. The word *if* is telling – technically, y is under a universal quantifier.

But we know by assumption that the implication $A \subseteq B$ is true. If we ever have the hypothesis of this implication, or a particular y, then we will know that its conclusion is true. Since we know that $x \in A$, we do have the hypothesis "$y \in A$" in the case $y = x$. The implication then tells that $x \in B$. You might think of this as *plugging in* the specific x into the general pattern that involved y.

Now that we have $x \in B$, the implication $B \subseteq C$ gives $x \in C$ with exactly the same reasoning as in the last two paragraphs. The statement $x \in C$ was what we were trying to prove: it's the implication of the conclusion $A \subseteq C$. We are done. □

In subsequent proofs, we won't be quite so detailed about the logic, trusting that you will pick up the general principles. But we will work to make sure you do pick them up!

A good introduction to doing mathematics is *How to read and do proofs*, by Daniel Solow (Wiley 1982). There are many books describing the use of set theory to build mathematics. *Naive Set Theory* by P. Halmos gives an informal introduction that leaves many details to the reader but does a good job explaining the purpose of the various axioms. The article *The Education of a Pure Mathematician*, by Bruce Pourciau in the American Mathematical Monthly, October 1999, uses a Socratic dialogue to introduce some of the philosophical questions involved in grounding mathematics. Beware that it is tricky to define a logical language for the entirety of mathematics; in fact, there are several ways to do that. The logic we use in our course is basic and void of controversy.

Problems

1. Let $\mathbf{A}, \mathbf{B}, \mathbf{C}$ be statements, each of which is either true or false Prove that this statement is always true.
$$(\mathbf{A} \Rightarrow \mathbf{B} \quad \text{and} \quad \mathbf{B} \Rightarrow \mathbf{C}) \quad \Rightarrow \quad (\mathbf{A} \Rightarrow \mathbf{C})$$
(Note: This is a problem of truth tables!)

2. For statements $\mathbf{A}, \mathbf{B}$, define the statement "$\mathbf{A}$ xor $\mathbf{B}$" to be true if $\mathbf{A}$ is true and $\mathbf{B}$ is false, and it is true if $\mathbf{A}$ is false and $\mathbf{B}$ is true. The statement $\mathbf{A}$ xor $\mathbf{B}$ is false otherwise. Show how to build this statement using *and, or, not*.

3. Let $\mathbf{A}, \mathbf{B}$ be statements. Show that
$$\Big(\mathbf{A} \Rightarrow \mathbf{B}\Big) \iff \Big[(\text{not } \mathbf{A}) \text{ or } \mathbf{B}\Big]$$

4. Let $\mathbf{A}, \mathbf{B}$ be statements. Show that the negation of "$\mathbf{A}$ and $\mathbf{B}$" is the statement "(not $\mathbf{A}$) or (not $\mathbf{B}$)."

5. Write elegant negations of the following statements.[5]

(a) All lions are fierce and there is a lion that is a vegetarian.

(b) I could not love thee, dear, so much, loved I not honor more.

(c) It is a far, far better thing that I do than I have ever done.

(d) Paris is beautiful every spring, or Paris was cold last winter, and Rome had a spring that was the same as Paris.

(e) Some courses are not as much fun as other courses.

6. Let A and B be sets.

(a) Formulate "A is a subset of B" using the universal quantifier.

(b) Formulate "A is not a subset of B" using the existential quantifier.

---

[5]*Elegance* precludes merely prefixing the word *not* to each statement!

7. Prove that $A \cap B = A$ implies that $A \subseteq B$. (As in the proof of Proposition 1.1, make sure you understand what you need to do and which symbols are specific objects and which are placeholders.)

8. Show that $A \subseteq B$ implies that $A \cap B = A$. (Note: the conclusion is that certain sets are *equal* – that means that each is a subset of the other, so prove that $A \cap B \subseteq A$ and $A \subseteq A \cap B$.)

9. Use the logic and set notation we have so far to formulate this statement: "S has exactly one element." Now try this one: "S has exactly two elements."

CHAPTER 2

The Integers

Now we are ready for the *integers*. This set, denoted $\mathbb{Z}$, can be constructed from very primitive axioms of logic and sets; we are more interested in initiating its formal study, and so we will introduce its properties axiomatically. Of course, intuitively,

$$\mathbb{Z} = \{\ldots -3, -2, -1, 0, 1, 2, 3, \ldots\}$$

and we will use the usual base 10 digit representation for specific integers. The curly brackets here are meant to enclose the elements of the *set* of integers. We will assume we know how to add and multiply specific integers in base 10 and how to compare two of them to see which one is larger.

We will begin with the formal properties of the integers we want to assume – our integer axioms. We have left out a very important property, induction, because it merits its own discussion. This list may appear long, but everything on it should be familiar. The important thing is this: a formal proof about the integers, given in class or on a homework set, may not use a fact other than those on the list, unless that fact has been proved (in class or by you). Right away we will see the need for additional facts; many will be developed during the coming class days.

AXIOMS 2.1. *Given integers a, b, there is an integer $a + b$ and an integer $a \cdot b$. Also, $a < b$ is a statement that is true or false. The following statements are true for every integer a, b, c:*

(1) $(a + b) + c = a + (b + c)$
(2) there is $0 \in \mathbb{Z}$ such that $a + 0 = a = 0 + a$
(3) there is $-a \in \mathbb{Z}$ such that $a + (-a) = 0 = (-a) + a$;
(4) $a + b = b + a$
(5) $a \cdot (b \cdot c) = (a \cdot b) \cdot c$
(6) there is $1 \in \mathbb{Z}$ such that $a \cdot 1 = a = 1 \cdot a$
(7) $a \cdot b = b \cdot a$
(8) $a \cdot (b + c) = (a \cdot b) + (a \cdot c)$ and $(a + b) \cdot c = (a \cdot c) + (b \cdot c)$
(9) we have $(-1) \cdot a = -a$;
(10) exactly one of the following is true: $a < b$, $a = b$, $b < a$;
(11) we have $a < (a + 1)$ and there are no integers between a and $a + 1$;
(12) if $a < b$ and $b < c$, then $a < c$;
(13) if $a < b$, then $(a + c) < (b + c)$;
(14) let $a < b$; if $c > 0$, then $a \cdot c < b \cdot c$.

We want to make sure you understand the need for the axioms. We stated the assumption that we can add specific integers. Thus, we can observe that $2 + 3 = 3 + 2$ and $10 + 13 = 13 + 10$. Statement (4) in the axioms says that $a + b = b + a$ for every $a, b \in \mathbb{Z}$. No matter how many *specific instances* of addition we observe, we will never see that *all of them* commute. It takes an axiom to go from specific observation to the pattern of a universal quantifier. We will discuss this further in class.

We will use all the usual integer notation. For example, we will write $a - b$ for $a + (-b)$, and we will sometimes write ab for $a \cdot b$. Regarding order, we will use the standard notations: "$x \leq y$" to mean "$x < y$ or $x = y$," and "$x \geq y$"

for "$x > y$ or $x = y$." We will also write $a > b$ for $b < a$. Here is an often used fact that follows from (10): if $x \leq y$ and $y \leq x$, then $x = y$.

Notice that our axioms do not define integer division; we will insist on avoiding fraction notation. It also does not define exponentiation x^n; we will attend to that later.

Not all of our axioms are strictly necessary; some of them can be derived as theorems from the others. In stating the axioms, we have tried to strike a balance between listing everything we could possibly need and being inconveniently brief. We will explore the logical connections among our axioms to some extent in class and on problem sets. Here are some additional properties of the integers; they follow directly from the axioms. We may prove some of them in class or leave some to homework. You will recognize property (6) for its importance in solving equations.

PROPOSITION 2.2. *Let $a, b, c \in \mathbb{Z}$. Then*

(1) if $a + b = 0$ then $b = -a$. (In other words, $-a$ is the only integer whose sum with a is 0.)
(2) $-(-a) = a$
(3) $a \cdot 0 = 0$
(4) if $a < b$, then $-a > -b$
(5) if $c < 0$ and $a < b$ then $c \cdot a > c \cdot b$
(6) if $a \cdot b = 0$, then $a = 0$ or $b = 0$
(7) if $a > 0$ then $a \geq 1$

Much of our attention will be on the set $\mathbb{N}$ of *positive* integers. Here is how this set is denoted.
$$\mathbb{N} = \{x \in \mathbb{Z} \mid x > 0\}$$
The curly brackets enclose a set; the vertical bar is read *such that*, and so the definition reads like this: "The set $\mathbb{N}$ is defined to be the set of all integers

that are positive." Thus, $\mathbb{N}$ is a subset of the integers: $\mathbb{N} \subseteq \mathbb{Z}$. We know that $0 < 1$, and so $1 \in \mathbb{N}$. Proposition 2.2(7) says that if $x \in \mathbb{N}$, then $x \geq 1$. It follows that 1 is the smallest (minimal) element of $\mathbb{N}$.

The set $\mathbb{N}$ is called the set of *natural numbers*. To complete our description of the integers, we need an additional axiom that relates to this set. This axiom of *well-ordering* is also related to induction. To state the axiom, we begin by saying that a *minimum* for a subset S of the integers is an element x of S such that $x \leq s$ for all $s \in S$. (Observe that $s = x$ is one case of this inequality.) The entire set of integers has no minimum element (prove this carefully from the axioms). Also, the empty set ϕ has no minimum. If a minimum exists, it must be unique, for if x, y are both minimums of S, then since $x \in S$ and y is a mimimum, we have $y \leq x$. Also, $y \in S$ and x is a minimum, so that $x \leq y$. We see that $x = y$; this proves that there can be only one minimum for a set of integers.

WELL-ORDERING. *If S is a non-empty subset of $\mathbb{N}$, then there is a minimal element of S.*

We can make this very plausible (we will do so in class). The fact that well-ordering needs to be an axiom rather than a theorem is a very profound fact that has been understood only since the late 1800's.

We have already remarked that 1 is the smallest element of $\mathbb{N}$. In particular examples of non-empty subsets S, the mimimum may or may not be obvious. The minimum positive integer that is even is 2; the minimum positive integer whose digits add up to 18 is 189. But what is the minimum positive integer consisting of five consecutive digits in the decimal expansion of π? Often the existence of a minimum is of use more abstractly than concretely.

You may have seen or done proofs *by induction*. We will present induction as a *property* of the natural numbers; we will see its utility in proofs extensively. As with well-ordering, we can make the induction axiom seem plausible.

INDUCTION. *Let T be a subset of $\mathbb{N}$, assume that $1 \in T$, and that if $x \in T$, then $x + 1 \in T$. Then $T = \mathbb{N}$.*

We will now show that well-ordering and induction are equivalent. This means that once you accept one of them, the other one follows as a theorem. Some people prefer well-ordering and some people prefer induction; the equivalence shows that it does not matter which one you prefer, you get both of them. We have called them both axioms, and we will feel free to use either of them in a given setting.

PROPOSITION 2.3. *Assume Axioms 2.1. Then well-ordering and induction are equivalent.*

PROOF. We proceed to establish that each of the axioms implies the other.

First, assume that well-ordering is true. In proving induction we are obliged to assume the hypothesis of that axiom. Thus we have a set $T \subseteq \mathbb{N}$ with $1 \in T$, and whenever $x \in T$ we have $x + 1 \in T$. We must show that $T = \mathbb{N}$. To do this, we define $S = \{y \in \mathbb{N} \mid y \notin T\}$.

Obviously, $T = \mathbb{N}$ if and only if $S = \{\}$. Assume that $S \neq \{\}$. Then well-ordering (which is assumed true) produces a minimal element y of S. Could $y = 1$? No, since by hypothesis $1 \in T$. Thus $y > 1$ (why?), so that $y - 1$ is a positive integer (why?). Since y is the minimal element of S, and since $y - 1 < y$, we must have $y - 1 \notin S$. Thus $y - 1 \in T$, but then the hypothesis on T shows that $y - 1 + 1 \in T$, in other words $y \in T$. This is a contradiction to the fact that $y \in S$.

The presence of a contradiction shows us that our most recent unforced assumption is false. That assumption was that S was non-empty. We conclude that S is empty, hence $T = \mathbb{N}$, as needed.

Now we assume induction, and we will prove the contrapositive of well-ordering: if S has no minimal element, then either S is not a subset of $\mathbb{N}$ or

S is empty. To prove this, we assume the hypothesis: that S has no smallest element. The conclusion we are proving is connected by the word *or*. If S is not a subset of $\mathbb{N}$, then the *or-statement* is true, and we are done. Thus, we can assume that $S \subseteq \mathbb{N}$, and we need to show that S is empty.

The proof that S is empty requires a subtle trick, define

$$T = \{x \in \mathbb{N} \mid y \in S \Rightarrow x < y\}$$

(In our proof we will go over very carefully the notation that defines the set T.) We will use induction to show that $T = \mathbb{N}$; it will follow that $S = \{\}$, for if $x \in S$, then $x \in \mathbb{N}$ implies that $x \in T$, whence the definition of T (with $y = x$) shows that $x < x$, which is impossible.

Now we show that $T = \mathbb{N}$. In the first place, $1 \in T$. Indeed, if $1 \notin T$, then the implication: $y \in S \Rightarrow 1 < y$ cannot hold (this is what it means for 1 *not* to be in T). That the implication is false provides us with a $y \in S$ such that 1 is not less than y, in other words, $y \leq 1$. But y is a natural number and 1 is the least natural number (do you remember why?). Thus $y = 1$. Now we got y from S, and we now see that $1 \in S$. But if $1 \in S$, then surely 1 is the smallest element of S, since S is a set of natural numbers, and 1 is the smallest natural number. That 1 is the smallest element of S contradicts the fact that S has no smallest element. This contradiction forces us to drop the idea that $1 \notin T$ (the assumption that started this paragraph). We conclude that $1 \in T$, after all.

Let $x \in T$, and we show that $x + 1 \in T$. If $x + 1 \notin T$, then (as in the last paragraph) there is $y \in S$ with $y \leq x + 1$. Since $x \in T$ we have $x < y$, so that Axiom 2.1(11) shows that $y = x + 1$, and therefore $x + 1 \in S$. We claim that $x + 1$ is the minimal element of S. Let $z \in S$; could z be less than $x + 1$? Since $x < z$ (because $x \in T$ and $z \in S$), we have that $z < x + 1$ is not allowed by Axiom 2.1(11). Thus $x + 1 \leq z$. We see that $x + 1$ is the minimal element for S, again a contradiction. Thus $x \in T$ forces $x + 1 \in T$.

The last two paragraphs, along with induction show that $T = \mathbb{N}$. As remarked above, this leads to S being empty, and this completes the proof. □

We now formally adopt well-ordering and induction as official axioms for the natural numbers, along with the Axioms 2.1. This completes the formal description of the integers. But don't forget that we also have Proposition 2.2 and anything else we can prove!

Here is an important use of induction: to make a general definition. Let's show that the notation 2^n makes sense. Informally, we think of 2^n as "multiplying 2 by itself n times," and if we have a specific n, we can compute the specific number. For instance $2^{11} = 2048$. But we want to define 2^n *for all* $n \in \mathbb{N}$, not just the ones we can write down as examples. That takes a proof! One of the uses of induction is to make informal ideas such as "multiply 2 by itself n times" into formal definitions.

Claim For each $n \in \mathbb{N}$ there is a unique number $2^n \in \mathbb{N}$, such that $2^1 = 2$, and $2^{n+1} = 2 \cdot 2^n$ for all $n \in \mathbb{N}$.

Solution. We define a set of positive integers that are exponents *up to a certain point*. Like this: Let T be the set of $k \in \mathbb{N}$ such that

(a) 2^n is defined for $n \in \mathbb{N}$ with $n \leq k$

(b) $2^1 = 2$

(c) for each $n \in \mathbb{N}$ with $n < k$, we have $2^{n+1} = 2 \cdot 2^n$

We will use induction to show that $T = \mathbb{N}$, and then statements (a), (b), (c) will show that 2^n is defined for all $n \in \mathbb{N}$ and with the desired properties. First, we need to see that $1 \in \mathbb{N}$; this is easy: $2^1 = 2$ establishes (a) and (b), and (c) is vacuous since there are no positive integers $n < 1$.

Now assume that $k \in T$ and we will show that $k+1 \in T$. Since $k \in T$, the number 2^k is defined. Define $2^{k+1} = 2 \cdot 2^k$. (This shouldn't surprise us at all – we want (c)!) Now we have 2^n defined for $n \leq k+1$ and we have $2^1 = 2$. For (c), let $n < k+1$. If $n < k$, then we have $2^{n+1} = 2 \cdot 2^n$ because (c) holds for k. If $n \geq k$, then $n = k$, and $2^{k+1} = 2 \cdot 2^k$ holds by the way we defined 2^{k+1}. We see that $k+1 \in T$.

By induction, $T = \mathbb{N}$, and we have the claim. ■

We can use exactly the same idea to define x^n for all $x \in \mathbb{Z}$ and $n \in \mathbb{N}$. Let's assume we've done this and use induction again to establish a familiar identity.

Claim Let $x \in \mathbb{Z}$ and $m, n \in \mathbb{N}$. Then $x^{m+n} = x^m \cdot x^n$.

Solution. Define T to be the set of $n \in \mathbb{N}$ such that $x^{m+n} = x^m \cdot x^n$ for all $x \in \mathbb{Z}$ and $m \in \mathbb{N}$. We will use induction to show that $T = \mathbb{N}$.

When $n = 1$, the desired identity is $x^{m+1} = x^m \cdot x^1$. From the definition of x^n (imitating the definition of 2^n), we know that $x^{m+1} = x^m \cdot x$. We also know that $x = x^1$, and we see that $1 \in T$.

Assume that $n \in T$, and we will show that $n+1 \in T$. Here is the calculation; make sure you understand *why each equation is true*.

$$x^{m+n+1} = x^{m+n} \cdot x = x^m \cdot x^n \cdot x = x^m \cdot x^{n+1}$$

We see that $n+1 \in T$. By induction, $T = \mathbb{N}$, and our formula holds in general. ■

Often, the use of induction is signaled by reference to a variable that runs over the natural numbers. For instance, the previous proof might start with the sentence, "Use induction on n." This sentence implies the more formal approach we used in the proof – involving the set T but without dealing with T directly.

Here is another use of induction: to derive an inequality.

Claim We have $2^n > n$ for each $n \in \mathbb{N}$.

Solution. Induction on n. (Be sure you understand how to define the formal set T to which induction is being applied.) We see that $2^1 = 2 > 1$, and so the

inequality holds for $n = 1$. Assume that $2^n > n$ for some particular $n \in \mathbb{N}$, and we compute
$$2^{n+1} = 2 \cdot 2^n > 2 \cdot n = n + n \geq n + 1$$
as needed. ∎

You will do a few representative proofs by induction to get used to it. It will also figure prominently in the more subtle proofs we will encounter in this course.

We need to extend well-ordering to bounded subsets of the integers. A subset S of the integers is *bounded above* if there is an integer x such that $s \leq x$ for all $s \in S$. The integer x is then an *upper bound* for S. If x is an upper bound, so is $x+1$ and $x+2$, etc., and so a set that is bounded above has many upper bounds. A *maximum* for S is an upper bound that is an element of S. A maximum would be unique, since if x, y are maximums, then since x is an upper bound and $y \in S$, we have $x \geq y$. But y is an upper bound and $x \in S$, and so $x \leq y$. Thus, $x = y$. Upper bounds are not unique; a maximum is unique.

The set S is *bounded below* if it has a *lower bound*: an integer x such that $x \leq s$ for all $s \in S$. A set that is bounded below has many lower bounds.

We will discuss the following fact in class, but you should think about why it is true before the discussion.

Claim Let S be a subset of the integers. Define T to be the set of $-x$ for all $x \in S$. If S is bounded below, then T is bounded above. If S is bounded above, then T is bounded below. ∎

Don't forget that there are many subsets of the integers that are neither bounded above nor bounded below: the entire set of integers, the set of even integers, etc.

The following result extends well-ordering to subsets of the integers that are bounded below or above. We will use the name *well-ordering* for the extended fact as well as for the original axiom. The proof will consist of two homework problems.

PROPOSITION 2.4. *Let S be a non-empty subset of the integers. If S is bounded below, then it has a minimal element. If S is bounded above, then it has a maximal element.*

Around 1860, Dedekind showed that integer addition and multiplication may be derived by assuming only four axioms. An excellent English translation of Dedekind's work is the small book *Essays on the Theory of Numbers* (Dover, 1963). These axioms were also employed by Peano in the 1880's as part of a particular way of building mathematics from the ground; thus, the axioms are often called *Peano's Axioms*. We will not make direct use of this approach, but we mention these four axioms, because it certainly is interesting that you can get all the properties of the integers from so little. Here they are.

(1) There is a set $\mathbb{N}$ (which will turn out to be the set of natural numbers), which has an element called 1.
(2) Given $a \in \mathbb{N}$ there is a "next element" $a + 1$ in $\mathbb{N}$, and $a + 1 \neq 1$.
(3) If $a, b \in \mathbb{N}$ and $a + 1 = b + 1$, then $a = b$.
(4) Induction holds.

Problems

10. Show that there is no integer x such that $17 \cdot x = 37$. (Note: remember that you are confined to the integer axioms; we do not have fractions! We can see that particular integers x are not solutions; to work in general, show that $17 \cdot x$ is either less than or greater than 37, depending on x.)

11. Show that there is no integer x such that $x^2 = 14$. (Remember to confine yourself to the axioms and what we have derived from them.)

12. Show that each of these integer axioms can be proved from the others.

(a) $a + b = b + a$ (Hint: think about $(a+b) \cdot (1+1)$.)

(b) $(-1) \cdot a = -a$ (Use that $a \cdot (1-1) = 0$.)

(c) $a \cdot b = b \cdot a$ (Hint: let $b \in \mathbb{Z}$ be given. Axiom (6) is that $1 \cdot b = b \cdot 1$. For the case $a \geq 1$, use induction on a. The cases $a = 0$ and $a \leq -1$ follow.)

13. In Axioms 2.1, delete axiom (14) and replace it with this: If $a > 0$ and $b > 0$, then $a \cdot b > 0$. Using this new axiom (14), prove the old (14): if $c > 0$ and $a > b$, then $c \cdot a > c \cdot b$.

14. Prove Proposition 2.2(6).

15. Prove the *Cancellation Rule*: if $a, b, c \in \mathbb{Z}$ and $a \cdot b = a \cdot c$, then either $a = 0$ or $b = c$.

16. Let S be a non-empty subset of the integers that is bounded below. Show that S has a minimum. (Hint: show that there is $b \in \mathbb{Z}$ such that $b < x$ for all $x \in S$. Look at the set of $x - b$ for all $x \in S$.)

17. Let S be a non-empty subset of the integers that is bounded above. Show that S has a maximum. (Hint: look back in the text for a method of turning S into a set that is bounded below.) Note that between this problem and the previous one, we have a proof of Proposition 2.4.

18. Use induction to give a formal proof of each of the following.

(a) $2 + 4 + 6 + \cdots + 2n = n \cdot (n+1)$ for all $n \in \mathbb{N}$

(b) $1 + 3 + 5 + \cdots + (2n-1) = n^2$ for all $n \in \mathbb{N}$

(c) $2^0 + 2^1 + \cdots + 2^n = 2^{n+1} - 1$

19. Let's show that addition on $\mathbb{N}$ can be used to define multiplication. We assume addition is defined on $\mathbb{N}$, that Axioms 2.1(1),(4) hold, and we assume induction. Show that for each $m, n \in \mathbb{N}$, there is a unique number $m \cdot n$ with

(a) $m \cdot 1 = m$

(b) $m \cdot (n+1) = (m \cdot n) + m$

(Hint: let m be given and use induction on n.) Once you have the definition, prove that

$$(a+b) \cdot n = (a \cdot n) + (b \cdot n) \quad \text{for all} \quad a, b, n \in \mathbb{N}$$

(Hint: induction on n.)

20. Prove that $(x \cdot y)^n = (x^n) \cdot (y^n)$ for all $x, y \in \mathbb{Z}$ and $n \in \mathbb{N}$.

21. Prove that $x^{m \cdot n} = (x^m)^n$ for all $x \in \mathbb{Z}$ and $m, n \in \mathbb{N}$. (Hint: the previous problem will prove useful at some point.)

22. Let $a, b \in \mathbb{Z}$ with $a, b \geq 0$. Prove that if $a < b$, then $a^2 < b^2$. Prove that if $a^2 < b^2$, then $a < b$. (Hint: the second statement follows from the first statement; how could $a \not< b$?)

23. This is a computational problem; show all work, writing out what you did to compute the answer, but you don't need much English explanation, if any. Find all integers $n \leq 50$ such that $n = a^2 + b^2$ for some integers a, b.

CHAPTER 3

Modular Arithmetic

The study of the integers is called *number theory*; in this chapter, we get number theory off the ground. Given integers m and n, we say that m *divides* n if there exists an integer k such that $m \cdot k = n$. We will almost always be concerned with the case where m and n are natural numbers. In this case, note that m divides n implies that $m \leq n$. Indeed, if $m \cdot k = n$, then k is positive, so that k is a natural number too. Then $k \geq 1$, and so, since $m > 0$ tells us that $m \cdot 1 \leq m \cdot k$, so that $m \leq n$. All the claims just made follow from the facts proved in Chapter 2. Check them!

Let us recall integer division with remainder.

DIVISION THEOREM. *Let m be a natural number, and let n be an integer. Then there are unique integers q (the* quotient*) and r (the* remainder*), such that $n = qm + r$ and $0 \leq r < m$.*

PROOF. The proof involves a clever use of well-ordering. Let

$$S = \{x \in \mathbb{Z} \mid 0 \leq x \text{ and } x = n - y \cdot m \text{ for some } y \in \mathbb{Z}\}$$

An informal way to describe S is to look at all integers:

$$\ldots n + 2 \cdot m\,,\ n + m\,,\ n\,,\ n - m\,,\ n - 2 \cdot m, \ldots$$

We collect the nonnegative integers in this list into the set S. Is anything in S, by the way? If $n \geq 0$, then $n = n - 0 \cdot m$ is an element of S. (Recall that $0 \cdot m = 0$ has been proved.) If $n < 0$, then note that $m \geq 1$ implies that $n \cdot (1 - m) \geq 0$. Thus $n \cdot (1 - m) = n - n \cdot m$ is in S.

The set S is nonempty, and it is bounded below (by 0 for instance). By well-ordering, S has a minimal element, call it r. Since $r \in S$, it can be written in the form $n - m \cdot q$ for some integer q. So far we have $n = m \cdot q + r$, with $0 \leq r$.

We claim that $r < m$. Otherwise, $r \geq m$, which tells us that $r - m \geq 0$. But also,

$$r - m = (n - m \cdot q) - m = n - m \cdot q - m = n - m \cdot (q+1)$$

These facts show that $r - m \in S$. Since $m > 0$, we see that $r - m < r$, and thus the minimality of r is contradicted. The contradiction stemmed from $r \geq m$, hence $r < m$, as claimed. Now $r \geq 0$, since $r \in S$, so that we have $0 \leq r < m$, as needed.

We must show that q and r are unique. Suppose that $n = m \cdot p + s$ where $0 \leq s < m$.

If r and s are different, then we have either $r < s$ or $s < r$. Assume, for example, that $s < r$. Keeping this in mind, compute

$$m \cdot q + r = n = m \cdot p + s$$

so that $\quad r - s = m \cdot (p - q)$

Since $s < r$, we have that $r - s$ is a natural number, and we see that m divides it. As we showed previously, it follows that $m \leq r - s$. But since $s \geq 0$, Axioms 2.1 shows that $r - s \leq r$. We also have $r < m$, and so now we have $r - s < m$, a contradiction. It must be that r and s are the same. Now $m \cdot q + r = n = m \cdot p + r$ (since $r = s$), and then $m \cdot q = m \cdot p$. Since $m \cdot q = m \cdot p$ and $m \neq 0$, the Cancellation Rule shows that $q = p$. This shows that q and r are unique. $\square$

Let $n \in \mathbb{Z}$, and take $m = 2$ as in the Division Theorem. Then we can write $n = 2 \cdot q + r$, where $r = 0$ or $r = 1$. Of course, the integers of the form $2 \cdot q$ are *even*, and those of the form $2 \cdot q + 1$ are *odd*.

We are ready to introduce the modular arithmetic. Our notation goes back to Gauss' influential and brilliant work *Disquisitiones Arithmeticae*.[1] Fix a natural number n; we call n the *modulus*. For integers a and b we write

$$a \equiv b \bmod n \quad \text{when} \quad n \text{ divides } a - b$$

The statement $a \equiv b \bmod n$ is read "a is *congruent* to $b \bmod n$." (The word "mod" is short for *modulo*.) When n is clear from context, as will almost always be the case, we write $a \equiv b$ and just say, "a is congruent to b." Here is a re-statement using the definition of *divides* more directly: $a \equiv b \bmod n$ means that there is an integer k such that $a - b = k \cdot n$

To gain familiarity with this concept, prove the following statements; they say that congruence works just like equality does. (We say that $\equiv$ is an *equivalence relation*.) Throughout, n, a, b, c are integers, and $n > 0$.

(1) $a \equiv a \bmod n$
(2) If $a \equiv b \bmod n$, then $b \equiv a \bmod n$
(3) If $a \equiv b \bmod n$, and $b \equiv c \bmod n$, then $a \equiv c \bmod n$.

We want to see what happens to integer arithmetic when "equality" is replaced by "congruence." For a natural number n as modulus, we define $\mathbb{Z}_n$ to be the integers, except that we do not distinguish between congruent integers. If $n = 4$, then since $3 \equiv 7$, the numbers 3 and 7 are the *same element* of $\mathbb{Z}_4$. (Of course, they are still distinct as elements of $\mathbb{Z}$.) We also have $4 \equiv 0$, and so 4 and 0 are the same element of $\mathbb{Z}_4$. In general in $\mathbb{Z}_n$, we regard n as being the same as 0 and play out the consequences in the other integers. When we do this, it is probably obvious that we are giving something away, for instance we can no longer use ordering: when $n = 5$ we have $2 \equiv 12$ rather than $2 < 12$. On the other hand, we will see that this point of view still allows us to do arithmetic.

---

[1]Gauss completed this work in 1801, and it has been published continuously since. Springer-Verlag published an accessible English version in 1986.

The Division Theorem shows that $\mathbb{Z}_3 = \{0, 1, 2\}$, the set of remainders when we divide by 3. That's because each integer m can be written $m = 3 \cdot q + r$, where $r \in \{0, 1, 2\}$, and notice that $m \equiv r$. In other words, $0, 1, 2$ are all the elements of $\mathbb{Z}_3$. More generally, let n be a positive integer modulus. If $x \in \mathbb{Z}$, then the Division Theorem finds $q, r \in \mathbb{Z}$ with $0 \leq r < n$ such that $x = q \cdot n + r$. We see that $x \equiv r$. In other words, every integer is congruent to a *remainder*. The uniqueness part of the Division Theorem shows that no two remainders are congruent to each other. Once we have discussed *counting* formally, we will see that this leads to the fact that $\mathbb{Z}_n$ has exactly n elements.

It is sometimes convenient to write $\mathbb{Z}_n = \{1, 2, \ldots, n\}$ rather than to use the remainders. This notation is more suggestive of what we said in the last paragraph: the set $\mathbb{Z}_n$ has n elements.

We like the informality of *replacing equality by congruence*. In later chapters we will encounter a more formal way to describe $\mathbb{Z}_n$. Some people use notation to distinguish between *integers* and elements of some $\mathbb{Z}_n$. Let's say we want to work with 5 as an element of $\mathbb{Z}_7$. We can write $\bar{5}$ for the element of $\mathbb{Z}_7$ to distinguish it from the ordinary integer 5. (There are other decorative notations sometimes used, as well.) If you think that such a notation will help you to keep $\mathbb{Z}$ and $\mathbb{Z}_n$ straight, feel free to use it.

The following Proposition will be used to define arithmetic in $\mathbb{Z}_n$.

PROPOSITION 3.1. *Let $n \in \mathbb{N}$ be the modulus. For integers a_1, a_2, b, suppose that we have $a_1 \equiv a_2 \mod n$. Then*

(a) $(a_1 + b) \equiv (a_2 + b) \mod n$
(b) $(a_1 \cdot b) \equiv (a_2 \cdot b) \mod n$

PROOF. We will take n for granted in writing the $\equiv$ symbol. If $a_1 \equiv a_2$, then $a_1 - a_2 = q \cdot n$ for some $q \in \mathbb{Z}$. Then

$$(a_1 + b) - (a_2 + b) = a_1 - a_2 = q \cdot n$$

This proves that $(a_1 + b) \equiv (a_2 + b)$. Also,
$$a_1 \cdot b - a_2 \cdot b = (a_1 - a_2) \cdot b = q \cdot n \cdot b$$
and this shows that $a_1 \cdot b \equiv a_2 \cdot b$. □

Looking at $\mathbb{Z}_6$, for example, consider the question whether $2 + 5$ makes sense. Of course, you say, since $2 + 5 = 7$. But we are using congruence in place of equality. Since $2 \equiv 8$, it should be possible to replace 2 by 8 in the expression $2 + 5$ and get the same expression in $\mathbb{Z}_6$ (not necessarily an equal expression in the integers). Proposition 3.1 looks at $2 \equiv 8$ and deduces that $2 + 5 \equiv 8 + 5$. In other words, 2 and 8 are interchangeable in addition, when congruence mod 6 is involved.

Proposition 3.1 shows that congruence can substitute for equality in expressions that involve addition and multiplication. Thus, addition and multiplication are defined in $\mathbb{Z}_n$. It is easy to see that strange and wonderful things follow. In $\mathbb{Z}_6$, for example, notice
$$2 \not\equiv 0 \quad \text{and} \quad 3 \not\equiv 0 \quad \text{but} \quad 2 \cdot 3 = 6 \equiv 0$$
In $\mathbb{Z}_5$, we have
$$2 \cdot 3 \equiv 1$$
and we are tempted to write
$$\frac{1}{2} \equiv 3$$
Once again we remind you that we do not have fractions, but it looks like we get them for free to some extent.

Because integer addition and multiplication are used to define addition and multiplication in $\mathbb{Z}_n$, the properties of the operations in $\mathbb{Z}_n$ are similar to those for the integers. We list the relevant properties, numbered to correspond to the integer Axioms 2.1.

PROPOSITION 3.2. *Fix a natural number modulus n, and let $a, b, c \in \mathbb{Z}_n$. Then we have the following.*

(1) $(a+b)+c \equiv a+(b+c)$;
(2) $a+0 \equiv a \equiv 0+a$;
(3) $a+(-a) \equiv 0 \equiv (-a)+a$;
(4) $a+b \equiv b+a$;
(5) $a \cdot (b \cdot c) \equiv (a \cdot b) \cdot c$;
(6) $a \cdot 1 \equiv a \equiv 1 \cdot a$;
(7) $a \cdot b \equiv b \cdot a$;
(8) $a \cdot (b+c) \equiv (a \cdot b) + (a \cdot c)$ and $(a+b) \cdot c \equiv (a \cdot c) + (b \cdot c)$.

Problems

24. Let n be a positive integer. Prove the following.

   (a) If $a, b \in \mathbb{Z}$ and $a \equiv b \mod n$, then $b \equiv a \mod n$.

   (b) If $a, b, c \in \mathbb{Z}$ and if $a \equiv b \mod n$ and $b \equiv c \mod n$, then $a \equiv c \mod n$.

25. If $m \in \mathbb{N}$, then there is no integer x such that $x^2 \equiv -1 \mod 4m$. (Hint: Is x even or odd?)

26. Suppose that $a, b, c \in \mathbb{Z}$ and $a^2 + b^2 = c^2$. Prove the following.

   (a) a, b cannot both be odd.

   (b) If c is even, both a, b are even.

   (c) If c is odd, then one of a, b is even and the other is odd.

(Hint: work in $\mathbb{Z}_4$.)

27. Find those $x \in \mathbb{Z}_{12}$ such that $x, 2x, 3x$, etc. gives all the elements of $\mathbb{Z}_{12}$. What do you notice about the x's you found?

28. Look at $2, 2^2, 2^3$, etc., in $\mathbb{Z}_{13}$. What interesting thing happens? How about the powers of 3? The powers of 4?

29. Which elements of $\mathbb{Z}_7$ can be written x^2 for some $x \in \mathbb{Z}_7$? Same question in $\mathbb{Z}_{11}$.

30. Find all integers $n \leq 30$ such that there is an integer x with
$$x^2 \equiv -1 \bmod n$$
(Show your work! Note that a previous problem rules out some of the n's.)

31. Consider the operation $a \star b = a^b$ for $a, b \in \mathbb{N}$. Is this an operation on $\mathbb{Z}_{13}$? Why, or why not?

CHAPTER 4

Functions

The notation $f : A \to B$ indicates that f is a *function* from the set A to the set B. This means that for each element a of A there is a uniquely determined element $f(a)$ of B. We call A the *domain* of f; alternatively we say that f is a function *on* A. A technicality: we allow A to be the empty set, even though $f : \phi \to B$ doesn't have any values $f(a)$, since ϕ has no elements. When we think of functions, we always imagine that the domain has elements.

We might imagine a function as a collection of "arrows" from the elements of A to elements of B. For each $a \in A$, there is one and only one arrow from a to an element of B (the element is $f(a)$). Notice that we can have $b \in B$ *not* at the end of an arrow, and we can have $b \in B$ at the end of several arrows.

Here is a picture depicting a very simple example. We have $A = \{p, q, r\}$ and $B = \{s, t, u\}$ with $f(p) = s = f(q)$, but $p \neq q$. Also, there is no $x \in A$ with $f(x) = t$.

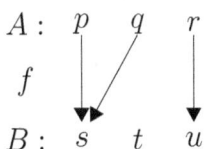

For functions f and g, we write $f = g$ and say that f and g are *equal* if they have the same domain, call it A, and if $f(a) = g(a)$ for every $a \in A$. In English: equal functions have the same domain and agree on where to send each element of that domain.

When $f : A \to B$, the set B is sometimes called the *range* of the function; we prefer not to use that term since it seems ambiguous. For instance, if B is a subset of C, and if $f : A \to B$, then we could say that $f : A \to C$ as well, and this would make both B and C "ranges" of f. We prefer to think about the *image* of the function f. If $f : A \to B$, then the image of f is the set of $b \in B$ for which there is $a \in A$ such that $f(a) = b$. Going back to the arrow picture; the image is the set of $b \in B$ at the ends of the arrows. We write $f(A)$ for the image of f. Here are two formal definitions:

$$f(A) = \{f(a) \mid a \in A\} = \{y \in B \mid \exists a (a \in A \text{ and } y = f(a))\}$$

The image of f is a subset of B. Since for $a \in A$ the element $f(a)$ is in $f(A)$ (by definition!), we will sometimes wish to consider f as a function from A to $f(A)$. Notice that equal functions have the same image.

Function composition occurs in algebra in a variety of settings. If

$$f : A \to B \quad \text{and} \quad g : B \to C$$

then we define the *composite function*

$$g \cdot f : A \to C$$

by the formula

$$(g \cdot f)(a) = g(f(a))$$

Notice that $f(a) \in B$ so that $g(f(a))$ makes sense and is an element of C. The act of forming $g \cdot f$ from f and g is called *function composition*.

In the composite notation $g \cdot f$, the function f is applied first, and then g is applied second. This "backwards" order is an unfortunate side effect of the notation $f(x)$ that writes the function f to the left of its argument x.

Believe it or not, we need to consider the case of three functions. The following is called the *associative law for functions*; its proof is very direct and will be given in class.

PROPOSITION 4.1. *Let $f : A \to B$, $g : B \to C$, and $h : C \to D$. Then $h \cdot (g \cdot f)$ and $(h \cdot g) \cdot f$ are functions from A to D and $h \cdot (g \cdot f) = (h \cdot g) \cdot f$.*

A useful but trivial example of a function is the *identity function* on the set A, which is denoted E_A and defined $E_A(a) = a$ for all $a \in A$. Thus, $E_A : A \to A$. It is easy to show, for $f : A \to B$, that

$$f \cdot E_A = f \quad \text{and} \quad E_B \cdot f = f$$

Notice that different identity functions are used on the two different sides of the function f; why is this? These equations are easy to verify once we remember what it means for functions to be equal.

Two more terms need be defined. Given $f : A \to B$, the image $f(A)$ of f is a subset of B; if this subset is all of B, then we say that f is *onto* B. For the sake of proofs, we re-formulate this: f is onto B if and only if for all $b \in B$ there is some $a \in A$ such that $f(a) = b$. Notice that this is **not** the same as the mere definition of a function. That $f : A \to B$ is that for each $a \in A$, there exists $f(a) \in B$. This starts with an arbitrary element of A; the definition of "onto" starts with an arbitrary element of B.

Again given $f : A \to B$, we say that f is *one to one* if $x, y \in A$ and $f(x) = f(y)$ imply that $x = y$. The (equivalent) contrapositive of this is often useful: if $x, y \in A$ and $x \neq y$, then $f(x) \neq f(y)$. Observe that the identity function E_A is both one to one and onto.

As with "onto", sometimes the definition of "one to one" is confused with the definition of function. A concrete example might illustrate this: define $f : \mathbb{Z} \to \mathbb{Z}$ by $f(x) = x^2$. If we pick, say, $3 \in A$, then there is a *unique* $b \in \mathbb{Z}$ such that $b = f(3)$, namely $b = 9$. The uniqueness of 9 is *not* that f is one to one, only that f is a function. For f to be one to one, the 3 would need to be unique: we would need to have $f(a) = 9$ implies that $a = 3$. But, of course, $f(-3) = 9$, as well, and so f is not one to one. When you prove that a function is one to one, make sure you prove that elements of the *domain* are unique, not elements of the image.

The following will be proved in class or on homework.

PROPOSITION 4.2. *Let $f : A \to B$ and $g : B \to C$.*
(a) If f and g are onto, then so is $g \cdot f$.
(b) If f and g are one to one, then so is $g \cdot f$.

Let $f : A \to B$ be onto. If $b \in B$, then there is some $a \in A$ such that $f(a) = b$ (the element a may not be unique!). Choose[1] such an a and define $g(b) = a$. Doing this for each $b \in B$ defines $g : B \to A$, such that $f(g(b)) = b$, in other words,

$$(4.1) \qquad f \cdot g = E_B$$

If f is not one to one, then there may be many functions g satisfying (4.1), because there may be many choices for a such that $f(a) = b$.

Conversely, assume there is a function $g : B \to A$ such that equation (4.1) holds. We can show that f is onto. Indeed, let $b \in B$, and then equation (4.1) says $f(g(b)) = E_B(b) = b$, so that $g(b)$ is an element of A which f sends to b. Thus f is onto.

Similarly, as you will prove, a function $f : A \to B$ is one to one if and only if there is a function $h : B \to A$ such that

$$h \cdot f = E_A$$

Let us put these concepts together. Let $f : A \to B$ be both one to one and onto. Then there are functions h and g from B to A such that

$$f \cdot g = E_B \quad \text{and} \quad h \cdot f = E_A$$

---

[1] If you are becoming especially acute in formalities, you will question what *choose* means. This is not a trivial issue; we take it up briefly at the end of this chapter.

Using Proposition 4.1 (the associative law),
$$h = h \cdot E_B = h \cdot (f \cdot g) = (h \cdot f) \cdot g = E_A \cdot g = g$$
so that g and h are the same function. Furthermore if k is a function from B to A such that either $k \cdot f = E_A$ or $f \cdot k = E_B$, then $k = g$, so that g and h are not only equal, but unique. We write f^{-1} for g (or h) in this case; f^{-1} is the *inverse* of f. Notice that the notation of the -1 exponent does not refer to division as it does in arithmetic.

We have shown that if $f : A \to B$ is both one to one and onto, then there is a unique function $f^{-1} : B \to A$ such that $f \cdot f^{-1} = E_B$ and $f^{-1} \cdot f = E_A$. Conversely, if $f : A \to B$ and there is a function $k : B \to A$ such that $k \cdot f = E_A$ and $f \cdot k = E_B$, then, by the previous paragraphs, f is both one to one and onto, so that $k = f^{-1}$. In particular, if f^{-1} exists, then the equations show that f^{-1} is itself one to one and onto, giving rise to a unique function $(f^{-1})^{-1} : A \to B$ such that $(f^{-1})^{-1} \cdot f^{-1} = E_B$ and $f^{-1} \cdot (f^{-1})^{-1} = E_A$. But the function f can be used in place of $(f^{-1})^{-1}$ in these equations, so that the uniqueness proved forces that $f = (f^{-1})^{-1}$. We summarize all of this.

PROPOSITION 4.3. *Let $f : A \to B$.*

(a) f is one to one if and only if there is $g : B \to A$ such that $g \cdot f = E_A$
(b) f is onto if and only if there is $g : B \to A$ such that $f \cdot g = E_B$
(c) f is one to one and onto if and only if there is a unique function
 $f^{-1} : B \to A$ *such that* $f \cdot f^{-1} = E_B$ *and* $f^{-1} \cdot f = E_A$

You have used one to one, onto functions many times. For example, when we count the elements of a set X and declare that it has 10 of them, we are setting up a one to one, onto function from the numbers $1, 2, \ldots, 10$ to the set X. The function is onto because each element of X is counted; the function is one to one because each element of X is counted exactly once.

We imagined using the integers $1, \ldots, 10$ to count to 10; these integers give us a model of a set with 10 elements. For each positive integer n, define $\mathbb{N}[n]$ to be the set of positive integers m such that $m \leq n$. This definition will be used only to discuss the properties of counting.

We define the set X to be *finite* if it is empty, or if there is a positive integer n and a one to one, onto function $f : \mathbb{N}[n] \to X$. In this context, we say that X has *order* n. We say that the empty set has order 0. For completeness, we present proofs of the fundamental facts about counting. In case we do not have time to discuss all of the proofs in class, we will be content to regard the facts as axioms. The main formal point is that counting follows from the basic properties of the natural numbers and functions.

The facts we are about to list should be familiar. For example, we prove that a finite set has a unique number of elements (you can't have a set of order 3 that also has order 7). Once this is proved, we can refer to that number as a function of the set being counted. For a finite set X, we will write $|X|$ for its order. Note that $|X|$ is *not* the absolute value notation in this context!

We need some set notation: for sets A, B their *union* $A \cup B$ is the set consisting of all x such that $x \in A$ or $x \in B$. The sets A and B are *disjoint* if $A \cap B$ is empty (if they have no elements in common).

PROPOSITION 4.4. *If X is a non-empty finite set, then there is a unique natural number n such that X has order n. In this situation, if X is the union of disjoint sets A and B, then A and B are finite and the order of X is equal to the sum of the order of A and the order of B.*

LEMMA 4.5. *Let $m \in \mathbb{N}$ and $a, b \in \mathbb{N}[m]$. Then there is $h : \mathbb{N}[m] \to \mathbb{N}[m]$, one to one and onto, such that $h(a) = b$.*

PROOF. Define $h : \mathbb{N}[m] \to \mathbb{N}[m]$ by
$$h(j) = \begin{cases} b & \text{if } j = a \\ a & \text{if } j = b \\ j & \text{if } j \notin \{a, b\} \end{cases}$$
Observe that the definition of h makes sense even when $a = b$. □

Proof of Proposition 4.4 Let T be the set of natural numbers n such that if a set has order n and also order m, then $n = m$. We prove that $T = \mathbb{N}$ by induction.

If X has order 1, then there is a one to one function f from $\mathbb{N}[1] = \{1\}$ onto X. It follows that $X = \{f(1)\}$. If g maps $1, \ldots, m$ one to one to X, and if $m > 1$, then $g(2) \neq g(1)$, yet both of these must be $f(1)$. This contradiction shows that $m = 1$, after all. Thus, $1 \in T$.

Assume that $n \in T$, and we will show that $n + 1 \in T$. To this end, let $f : \mathbb{N}[n+1] \to X$ be one to one and onto, and let $g : \mathbb{N}[m] \to X$ be one to one and onto. We need to prove that $n = m$. If $m = 1$, then $n + 1 = 1$ by the fact that $1 \in T$. It follows that $n = 0$, a contradiction. Thus, $m > 1$.

We want to modify g so that $g(m) = f(n+1)$. Since $f(n+1) \in X$ and g is onto, there is $j \in \mathbb{N}[m]$ such that $g(j) = f(n+1)$. We apply Lemma 4.5 with $a = m$ and $b = j$; that lemma finds $h : \mathbb{N}[m] \to \mathbb{N}[m]$ such that $h(m) = j$. Since h and g are one to one and onto, Proposition 4.2 shows that their composite $g \cdot h$ is one to one and onto. Compute that $g \cdot h(m) = g(j) = f(n+1)$. We replace g by $g \cdot h$ and now we have $g(m) = f(n+1)$.

Define X' to be the set of elements of X not equal to $f(n+1)$, and it is easy to see that f maps $\mathbb{N}[n]$ one to one, onto X', and $g : \mathbb{N}[m-1] \to X'$ is one to one, onto, as well. Since $n \in T$, we have that n is unique, and so $n = m - 1$. It follows that $n + 1 = m$, as needed. This proves that $n + 1 \in T$.

We conclude that $T = \mathbb{N}$, and so the order of a finite set is unique. For the rest of the proof, we use the notation $|X|$ for the order of X.

For the second statement, we again use induction. Let T be the set of natural numbers n such that if X has order n, and if $X = A \cup B$ with $A \cap B = \phi$, then A and B are finite and $|X| = |A| + |B|$. If $|X| = 1$, then the hypothesis would give $X = A$ and $B = \phi$ or $X = B$ and $A = \phi$. Thus, $1 \in T$.

Now assume that $n \in T$ and $|X| = n+1$, and $X = A \cup B$, as above. Let $f : \mathbb{N}[n+1] \to X$ be one to one and onto. Let $x = f(n+1)$. Define X' to be the set of elements of X not equal to x, and we can use the function f to see that $|X| = |X'| + 1$.

Now we know that $x \in A$ or $x \in B$, and not both. Say, for example, that $x \in B$. Define B' to be the set of elements of B not equal to x, and then X' is the disjoint union of A and B'.

Recall the function f. We have $f : \mathbb{N}[n] \to X'$ is one to one and onto, and so, since $n \in T$, we conclude that A and B' are finite, and

$$|X'| = |A| + |B'|$$

It is easy to see that $|B| = |B'| + 1$. Indeed, if $B' = \phi$, then $B = \{x\}$ has order 1. Otherwise, if $|B'| = m \geq 1$, then there is $g : \mathbb{N}[m] \to B'$, one to one and onto. Defining $g(m+1) = x$, we see that $g : \mathbb{N}[m+1] \to B$ is one to one and onto. Thus, $|B| = |B'| + 1$, as claimed. Now

$$|X| = |X'| + 1 = |A| + |B'| + 1 = |A| + |B|$$

∎

Here are some more elementary counting principles. Proposition 4.6a is often called the *pigeon-hole principle*; this term is explained by imagining the set A as counting nests (pigeon-holes) and B as counting nests that have a pigeon in them.

PROPOSITION 4.6. *Let A be a finite set.*

(a) If B is a subset of A, then B is finite and $|B| \leq |A|$. If, furthermore, $|B| = |A|$, then $B = A$.

(b) If $f : A \to B$ is one to one and onto, then B is finite, and $|A| = |B|$.

(c) If $f : B \to A$ is one to one, then B is finite, and $|B| \leq |A|$.

(d) If $f : A \to B$ is onto, then B is finite and $|B| \leq |A|$.

PROOF. For (a), define C to be the set of $x \in A$ such that $x \notin B$. Then $A = B \cup C$, and B and C are disjoint. By Proposition 4.4, we have that B is finite and that $|A| = |B| + |C|$, so that $|B| \leq |A|$. If $|B| = |A|$, then obviously, $|C| = 0$, so that C is empty.

For (b), let $h : \mathbb{N}[n] \to A$ be one to one and onto, and look at $f \cdot h$.

For (c), notice that $f(B)$ is a subset of A. Apply (a) and (b).

For (d), there is $g : B \to A$ that is one to one. □

We introduce another important construction. For sets A and B, let $a \in A$ and $b \in B$, and we can form the *ordered pair* (a, b) with coordinates a and b. Ordered pairs (a_1, b_1) and (a_2, b_2) are *equal* if $a_1 = a_2$ and $b_1 = b_2$. The set of all ordered pairs (a, b) with $a \in A$ and $b \in B$ is denoted $A \times B$; this set is the *Cartesian product of A and B*.

If either A or B is empty, then $A \times B$ is empty as well. We will see that the order of a Cartesian product of finite sets is the product of the orders of the sets. To prove this we need a lemma that will be completed in class or by you.

LEMMA 4.7. *Let $m, n \in \mathbb{N}$. Then there is a one to one, onto function $q : \mathbb{N}[m] \times \mathbb{N}[n] \to \mathbb{N}[m \cdot n]$.*

PROOF. For $i \in \mathbb{N}[m]$ and $j \in \mathbb{N}[n]$, define $q(i, j) = i + m \cdot (j - 1)$. □

PROPOSITION 4.8. *Let A and B be finite sets. Then $A \times B$ is finite, and $|A \times B| = |A| \cdot |B|$.*

PROOF. If A is empty, then $A \times B$ is empty, and the formula we are trying to prove reads $0 = 0 \cdot |B|$. Similarly, we are done if B is empty.

Assuming that A and B are non-empty, let $|A| = m$ and $|B| = n$, where $m, n \in \mathbb{N}$, and get one to one, onto functions $f : \mathbb{N}[m] \to A$ and $g : \mathbb{N}[n] \to B$. We can define $h : \mathbb{N}[m] \times \mathbb{N}[n] \to A \times B$ by

$$h(i,j) = (f(i), g(j))$$

It is easy to see that h is one to one, onto. Lemma 4.7 gives us a function $q : \mathbb{N}[m] \times \mathbb{N}[n] \to \mathbb{N}[m \cdot n]$ that is one to one, onto, and the composite function $h \cdot q^{-1}$ then maps $\mathbb{N}[m \cdot n]$ one to one, onto $A \times B$. □

Here is another use of functions and Cartesian products.. An *operation* $\circ$ on a set A is a way of producing an element $x \circ y$ of A from given elements x and y of A (x and y do not have to be distinct). You know many examples: addition and multiplication are operations on the set of integer they are also operation on the set of real numbers; division is an operation on the set of non-zero real numbers; matrix addition and multiplication are operations on the set of 2 by 2 matrices; vector addition is an operation on the set of vectors in 3-space. An operation $\circ$ on the set A is just a function from $A \times A$ to A, but we usually write $x \circ y$ instead of the more explicit function notation $\circ(x, y)$.

Let A be a set, and let F be the set of functions $f : A \to A$. Then function composition is an operation on F. This example will occur over and over in the course.

A function can be viewed as a subset of a Cartesian product; indeed, this is familiar to you from graphing functions in the xy-plane. If $f : A \to B$, we can define

$$C = \{(a, f(a)) \mid a \in A\}$$

The set C is a subset of $A \times B$; it is called the *graph* of f. If we are given C, we can recover the function f; given $a \in A$, there is only one $b \in B$ such that $(a, b) \in C$, and we have $b = f(a)$.

It is easy to see that not all subsets of $A \times B$ are graphs. Let $A = B = \mathbb{Z}$, and let $C = \{(1, 1), (1, 2)\}$, and we see that we can't define $f(1)$, because it is ambiguous. If $C \subseteq A \times B$, then C is the graph of a function if and only if for every $a \in A$, there is exactly one $b \in B$ such that $(a, b) \in C$.

We end with two technical matters. First, a matter of terminology. Around the 1940's an attempt was made to standardize mathematical terms. This attempt involved a particular grounding of mathematics. The group of mathematicians who developed this approach referred to themselves by the pseudonym "N. Bourbaki." They preferred to say that a function is *injective* rather than that it is one to one. An injective function may be called an *injection*. An onto function is *surjective*; it is a *surjection*. A one to one and onto function is *bijective*; it is a *bijection*. Although we will not use these terms in class, you will find them in much of the literature, and some otherwise very able instructors will use them. An extensive discussion of terminology that originated in Bourbaki is found in *Categories for the Working Mathematician* by S. MacLane (Springer-Verlag, 1971).

Second, when we had an onto function $f : A \to B$ and constructed a function $g : B \to A$ such that $f \cdot g = E_B$, we called attention to (and tried to cast suspicion on) the use of the word *choose*. Look back at the construction if you don't remember what was chosen. Choosing elements from the sets we will consider in math 300 can be justified by induction arguments; choosing elements from *arbitrary* sets involves an axiom of set theory that we have not discussed, the *Axiom of Choice*. This somewhat controversial axiom can be hinted at by saying that it is a kind of "super-induction." The full-blown Axiom of Choice will not be needed in math 300, and so we will not go into more detail. The book *Naive Set Theory*, by Halmos (referred to at the end of Chapter 1), gives an interesting introduction to this axiom. A more advanced and more complete treatment occurs in *Equivalents of the Axiom of Choice*, by H. Rubin and J. Rubin, North Holland Publishing, 1970.

Problems

32. Let $f : \{a, b, c, d, e\} \to B$ where $B = \{1, 2, 3\}$ and $f(a) = f(b) = 1$ and $f(c) = f(d) = 2$ and $f(e) = 3$. Find all functions $g : B \to A$ such that $f \cdot g = E_B$.

33. Let $f : A \to B$. Show that f is one to one if and only there is $g : B \to A$ such that $g \cdot f = E_A$.

34. Find a specific function $f : \mathbb{N} \to \mathbb{N}$ that is one to one and not onto. For a solution, give an explicit recipe for the function and then a formal proof that it is one to one and that it is not onto.

35. Find a specific function $g : \mathbb{N} \to \mathbb{N}$ that is onto and not one to one.

36. Prove Proposition 4.2b.

37. Let a, b be integers with $a \leq b$. Define $[a, b]$ to be the set of integers x with $a \leq x \leq b$. Prove that $[a, b]$ is finite by constructing a counting function $f : \mathbb{N}[n] \to [a, b]$ explicitly. (Hint: what's n?)

38. Let $X \subseteq \mathbb{Z}$ and suppose that X has a minimum and a maximum. Show that X is finite. (Hint: use the previous problem along with Proposition 4.6.)

39. Let X be a non-empty, finite subset of the integers. Show that X has a minimum and a maximum. (Hint: induction on $|X|$.)

40. Let S be a finite, non-empty set of integers. Show that S can be *put into order*. This means that if $n = |S|$, then there is a function

$$f : \mathbb{N}[n] \to S$$

such that $f(1) < f(2) < \cdots < f(n)$. Make sure your proof is formal! (Hint: induction on n.)

41. An *infinite set* is a set that is not finite. Let S be an infinite set. Then there is $f : \mathbb{N} \to S$ that is one to one. (Hint: by induction, show that there are distinct elements $f(1), f(2), \ldots, f(n)$ of S.)

42. Let A, B, C, D be sets and suppose there is a one to one, onto function $f : A \to B$ and a one to one, onto function $g : C \to D$. Show that there is a one to one, onto function $h : A \times C \to B \times D$.

43. Give a formal proof that a set of order n has exactly 2^n subsets. (Note: recall that we constructed 2^n on p.17.)

44. Let A and B be non-empty finite sets. Let F be the set of functions $f : A \to B$. Show that F is finite and $|F| = |B|^{|A|}$. (Hint: induction on $|A|$.)

45. Suppose that S is a set of statements, each of which is true or false. Define $f : S \to \mathbb{Z}_2$ by $f(A) \equiv 1$ if A is true, and $f(A) \equiv 0$ if A is false, for all $A \in S$. Prove the following, in which all congruences are mod 2 and in which S contains A and B and all the other statements referred to in (a)-(c).

  (a) $f(A \text{ and } B) \equiv f(A) \cdot f(B)$
  (b) $f(A \text{ or } B) \equiv f(A) \cdot f(B) + f(A) + f(B)$
  (c) $f(\text{not } A) \equiv 1 + f(A)$

Express $f(A \Rightarrow B)$ in terms of $f(A)$ and $f(B)$ and arithmetic in $\mathbb{Z}_2$. Express $f(A \text{ xor } B)$ in terms of $f(A)$ and $f(B)$ and arithmetic in $\mathbb{Z}_2$. (The logical *xor* is defined in a problem on p.9.)

46. Let C be the graph of $f : A \to B$. How can we tell from the set C that f is one to one? That f is onto? Your answers should be in terms of the set C, without mentioning the function f.

47. The *binomial coefficients* $\binom{n}{k}$ are defined for all non-negative integers $n \geq k$. We have the following formulas.

$$\binom{n}{0} = \binom{n}{n} = 1$$

$$\binom{n}{k} = \binom{n-1}{k-1} + \binom{n-1}{k} \quad \text{when} \quad 1 \leq k < n$$

Prove that

$$n! \cdot k! \cdot \binom{n}{k} = n! \quad \text{for all} \quad 0 \leq k \leq n$$

(Hint: induction on n. Remember, we don't have fractions!)

48. For $n, k \in \mathbb{N}$, define $d(n, k)$ to be the number of *onto* functions

$$f : \mathbb{N}[n] \to \mathbb{N}[k]$$

(a) Show that

$$d(n, 1) = 1 \quad \text{and} \quad d(n, k) = 0 \quad \text{if} \quad n < k$$

(b) For $n \geq 2$ and $k \geq 2$, show that

$$d(n, k) = k \cdot d(n-1, k) + k \cdot d(n-1, k-1)$$

(Hint: think about what f does to $N[n-1]$; is it onto?)

(c) Find $d(5, 3)$.

49. Let $f : \mathbb{Z} \to A$. Choose $n \in \mathbb{N}$. If $a \equiv b \bmod n$ implies that $f(a) = f(b)$, for every $a, b \in \mathbb{Z}$, then we can think of f as a function from $\mathbb{Z}_n$ to A. As an example, let $f : \mathbb{Z} \to \mathbb{Z}$ be defined by saying that $f(m)$ is the remainder when m is divided by 3.

(a) Does f define a function from $\mathbb{Z}_6$ to $\mathbb{Z}$?

(b) Does f define a function from $\mathbb{Z}_5$ to $\mathbb{Z}$?

CHAPTER 5

Permutations

A *permutation* on the set P is a one to one function from P onto itself. Permutations will be a central object of study in this course. There are several standard notations for the set of all permutations on a set P; we will use $\mathbb{S}_P$. Following time-honored usage, we will refer to the elements of P as *points* in this context. There is a special case where the notation varies; let n be a natural number, and suppose that P is the set of natural numbers k with $k \leq n$. We will write $\mathbb{S}_n$ for $\mathbb{S}_P$ in this case. (In Chapter 4 we used the notation $N[n]$ for the set of natural numbers less than or equal to n; we will not persist with that notation, since it is not standard.)

Given a set P, the identity function E_P is a permutation ($E_P \in \mathbb{S}_P$). If P has just one element, then E_P is the only permutation of P. If $P = \{a, b\}$, then there are two permutations: E_P and the function f such that $f(a) = b$ and $f(b) = a$. You will show that a finite set of order n has exactly $n!$ permutations.

If we are given $f : P \to P$, and we wish to prove that f is a permutation, the definition seems to force us to prove both that f is onto and that it is one to one. In general, this is exactly what we must do, but if P is a finite set, the following tells us that it is sufficient merely to check one of these properties.

PROPOSITION 5.1. *Let P be a finite set, and let $f : P \to P$. If f is either one to one or onto, then it is a permutation.*

PROOF. Assume that f is one to one. Then f defines a function from P to $f(P)$, and this function is both one to one and onto. Proposition 4.6 forces that $|P| = |f(P)|$. On the other hand $f(P) \subseteq P$, and so the same proposition shows that $P = f(P)$. Thus, $f : P \to P$ is onto, so that f is a permutation.

Now assume that f is onto. Then there is a one to one function $g : P \to P$ such that $f \cdot g = E_P$. The previous argument showed that g is onto, and so it has an inverse. Since $f \cdot g = E_P$, we have $f = g^{-1}$, and so f is one to one. $\square$

Since we are going to work with permutations all the time, we need a good notation for them. To get this notation we introduce a special kind of permutation: the *cycle*. Suppose, for example, that a, b, c are distinct elements of P. Then the cycle $[abc]$ is the permutation that behaves like this, using parentheses for function notation:

$$[abc](a) = b$$
$$[abc](b) = c$$
$$[abc](c) = a$$
$$[abc](x) = x \quad \text{if} \quad x \in P \setminus \{a, b, c\}$$

The permutation $[abc]$ "cycles" a to b, b to c and c to a, and it *fixes* (leaves alone) all other points (all other elements of P).[1] Notice that $[abc] = [bca] = [cab]$, so that any particular point in a cycle can be written first. We call $[abc]$ a *3-cycle*.

Here is a general, necessarily somewhat abstract, definition of a cycle. Let

$$a_1, a_2, \ldots, a_k$$

be distinct elements of P. The cycle $[a_1 a_2 \ldots a_k]$ is the function from P to P which sends a_i to a_{i+1} for $1 \leq i \leq k-1$, sends a_k to a_1, and fixes every other point (sends every other point to itself). It is easy to see that a cycle is a permutation – we will have a nice formula for the inverse presently. A cycle involving k points is a *k-cycle*.

---

[1] Points that are fixed are called *fixed points*.

Here is another example, albeit simpleminded. For $a \in P$, the cycle $[a]$ is the identity element, $[a] = E_P$.

Suppose that $|P| = n$. If k is a positive integer, how many permutations in $\mathbb{S}_P$ are k-cycles?

The main point of this chapter is that every permutation is a product (composite) of cycles. We need to prove this and to show how to find the cycles for a given permutation. Calculations on permutations are most easily done using the cycles; we will do examples in class, but you will need to work to master the details.

The cycles $[a_1 \cdots a_k]$ and $[b_1 \cdots b_j]$ are *disjoint* if the sets of points involved are disjoint:

$$\{a_1, \ldots, a_k\} \cap \{b_1, \ldots, b_j\} = \phi$$

PROPOSITION 5.2. *Disjoint cycles commute in function composition. The inverse of the cycle $[a_1 \cdots a_k]$ is the cycle $[a_k \cdots a_1]$.*

PROOF. Let $a = [a_1 \cdots a_k]$ and $b = [b_1 \cdots b_j]$ be disjoint cycles, and let $x \in P$. If x is one of the a_i, then $b(x) = x$ (since the cycles are disjoint!), and $a(x)$ is one of the a_i, so that $b(a(x)) = a(x)$, as well. Thus,

$$(a \cdot b)(x) = a(b(x)) = a(x) \quad \text{and} \quad (b \cdot a)(x) = b(a(x)) = a(x)$$

Similarly, if x is one of the b_i, then $a \cdot b$ and $b \cdot a$ agree on x. In the remaining case, $b(x) = x$ and $a(x) = x$, so that $a \cdot b$ and $b \cdot a$ send x to x. Thus, $a \cdot b = b \cdot a$.

The claim about inverses is left to you. □

Our main result: every permutation is a product of disjoint cycles. The proof is not all that hard, but there is a lot of bookkeeping involved. To anticipate the general argument, consider the following specific example. We consider the element f of $\mathbb{S}_{11}$ defined by writing $f(x)$ below x in the following table, for $1 \leq x \leq 11$.

$$\begin{array}{c|ccccccccccc} x: & 1 & 2 & 3 & 4 & 5 & 6 & 7 & 8 & 9 & 10 & 11 \\ f(x): & 9 & 10 & 6 & 2 & 5 & 11 & 3 & 8 & 1 & 4 & 7 \end{array}$$

We choose the point 3 rather arbitrarily, and we apply f over and over: $f(3) = 6$ and $f(6) = 11$ and $f(11) = 7$ and $f(7) = 3$. We see that we end up at the point 3 where we started. This shows that f involves the cycle $[3, 6, 11, 7]$ in the sense that this cycle shows what f does to the points involved in the cycle. Choosing another point, say 2, we see that $f(2) = 10$ and $f(10) = 4$ and $f(4) = 2$. Again, we get back to the point where we started. The cycle $[2, 10, 4]$ shows what f does to the three points in this cycle. You should see that the cycle $[1, 9]$ shows what f does to 1 and 9. Also, the points 5 and 8 are fixed by f, and so f involves the cycles $[5]$ and $[8]$. If we put all these cycles together, we get the entire function f:

$$f = [3, 6, 11, 7] \cdot [2, 10, 4] \cdot [1, 9] \cdot [5] \cdot [8]$$

Remember that the operation here is function composition. It is easy to check that this equation is valid. The unique cycle where a given point is mentioned shows what f does to that point.

By Proposition 5.2, the disjoint cycles making up f commute, and so they can be written in any order we want. Now we will prove that the process just carried out writes every permutation as a product of disjoint cycles.

PROPOSITION 5.3. *Let f be a permutation on the finite set P. Then f is the product of pairwise disjoint cycles. The sum of the lengths of all the cycles is $|P|$.*

PROOF. Let $i \in P$, and we will show how to find a cycle for f that involves i. Consider the sequence

$$i, \quad f(i), \quad f(f(i)), \quad f(f(f(i))), \quad \ldots$$

Let's write $i = f^0(i)$, and $f(i) = f^1(i)$ and $f(f(i)) = f^2(i)$, and so on, so that our list has the form $f^j(i)$ for $0 \leq j \in \mathbb{Z}$.

Since P is finite, this list must have repeats in it. Let us find the *first* repeat, that is to say the first k such that $f^k(i)$ has already occurred among the $f^j(i)$ with $j < k$. (Stop and answer the question, "Why does k exist?") We will show that this first repeat is i: the point we started with.

Find some particular $j < k$ such that

$$f^j(i) = f^k(i)$$

What we claim is that $j = 0$. Suppose that $j > 0$, the only other alternative, and apply f^{-1} to both sides of the equation to get

$$f^{-1}(f^j(i)) = f^{-1}(f^k(i)) \quad \text{so that}$$
$$f^{j-1}(i) = f^{k-1}(i)$$

Since $1 \leq j < k$, we see that $0 \leq j - 1 < k - 1$, so that the points $f^{j-1}(i)$ and $f^{k-1}(i)$ occur in the sequence of points. But the equation then says that $f^{k-1}(i)$ is a repeat, and this contradicts the choice of k as first. Thus $j = 0$, and we have $f^k(i) = f^0(i) = i$.

This shows that the function f agrees with the cycle

$$[i \; f(i) \; f^2(i) \ldots f^{k-1}(i)]$$

on the points in the list. We give this cycle the temporary name $c(i)$.

Now begin all over again with $j \in P$ such that j does not occur in the list for $c(i)$. The same argument produces a cycle $c(j)$ that gives the action of f on j and $f(j)$, etc. We claim that $c(i)$ and $c(j)$ are disjoint. Indeed, if p is a point occurring in the cycle $c(i)$, then $f^m(p)$ is in the cycle $c(i)$ for all $m \geq 0$. If the point p is also in the cycle $c(j)$, then $f^m(p)$ is also in $c(j)$. Furthermore, some power of m will give $f^m(p) = j$, and now we see that j occurs in the cycle $c(i)$, a contradiction.

Repeating the search for cycles over and over, we arrive at a finite number of disjoint cycles such that every element of P occurs in exactly one cycle. If $p \in P$, then there is a unique cycle $c(q)$ in which p appears. In the product of the disjoint cycles, only $c(q)$ does not send p to itself, and $c(q)$ sends p to $f(p)$, another point in the same cycle. This shows that the product of the cycles is equal to f as a function. Since every point is mentioned exactly once among the disjoint cycles, the sum of the cycle lengths is $|P|$. $\square$

Proposition 5.3 gives us the normal way of writing a permutation. You should try to write all the elements of $\mathbb{S}_4$ (the permutations of $1, 2, 3, 4$), as products of cycles. We will complete the work in class.

There are times when we wish to write a permutation as a product of a particular kind of cycle. Here is an example. Note well that it does not refer to disjoint cycles.

PROPOSITION 5.4. *Let $f \in \mathbb{S}_P$ and suppose that P is finite with at least 2 elements. Then f is a product of 2-cycles.*

PROOF. By Proposition 5.3 it suffices to show that a cycle is a product of 2-cycles. For example if we have a cycle with a single point $[a]$, then the fact that $|P| \geq 2$ produces an element b of P, distinct from a, and then $[ab]$ is a 2-cycle. Since $[a] = [ab][ab]$, we see that $[a]$ is a product of 2-cycles.

For a cycle $[i_1 i_2 \ldots i_r]$ with $r \geq 2$, compute

$$[i_1 i_2 \ldots i_r] = [i_1 i_2][i_2 i_3] \cdots [i_{r-1} i_r]$$

$\square$

The 2-cycles of Proposition 5.4 do not have to be disjoint, and, given $f \in \mathbb{S}_P$, there may be many ways to write f as a product of 2-cycles. For example

$$[123] = [13][12] = [23][13] = [12][13][45][23][12][54]$$

We see that the 2-cycles are not unique; in fact the number of 2-cycles used is not unique. However, for a given permutation, the number is always even or always odd, as we will prove. This is a very important fact with many applications. To get started, we need to show how to move one 2-cycle across a product of others. The main point of the following is that when a 2-cycle moves across, it preserves the *number* of 2-cycles it crosses.

PROPOSITION 5.5. *Let P be a non-empty set, and let $x, y \in \mathbb{S}_P$, where x is a 2-cycle and y is the product of k 2-cycles. Then there is a product y' of k 2-cycles such that $x \cdot y = y' \cdot x$.*

PROOF. Write $x = [ab]$, and it suffices to show how x moves across a single 2-cycle $[cd]$. We consider the possible overlap between points a, b, c, d. Of course, we have $a \neq b$ and $c \neq d$. Here are the possibilities:

$$[ab] \cdot [ab] = [ab] \cdot [ab]$$
$$[ab] \cdot [ad] = [bd] \cdot [ab] \qquad \text{if } b \neq d$$
$$[ab] \cdot [bc] = [ac] \cdot [ab] \qquad \text{if } a \neq c$$
$$[ab] \cdot [cd] = [cd] \cdot [ab] \qquad \text{if } \{a,b\} \cap \{c,d\} = \phi$$

□

PROPOSITION 5.6. *Let P be a finite set with at least 2 elements, and suppose we can write the identity permutation $E = E_P$ as a product of k 2-cycles in $\mathbb{S}_P$. Then k is even.*

PROOF. Induction on k. The case $k = 1$ is simply that a single 2-cycle is not E. (We are saying that when $k = 1$, the implication is vacuous.)

Let $x_1, \ldots, x_k$ be 2-cycles, with $k \geq 2$ and

$$E = x_1 \cdot x_2 \cdots x_k$$

We will show that E can be written as a product of $(k-2)$ 2-cycles.

Write $x_k = [ab]$, and we consider the 2-cycles that involve a. Since $E(a) = a$, the 2-cycle x_k cannot be the only one of the x_j involving a. Let $x_j = [ac]$ for some $j < k$.

We move x_j to the right, next to x_k. To do this, we quote Proposition 5.5 to show that

$$x_j \cdot x_{j+1} \cdots x_k = x'_{j+1} \cdots x'_{k-1} \cdot x_j \cdot x_k$$

where x'_i are 2-cycles, for $j+1 \leq i \leq k-1$. Substituting this into the expression for E, we see that E is a product of k 2-cycles, where the rightmost 2-cycles are $x_j \cdot x_k$.

Remembering that $x_j = [ac]$ and $x_k = [ab]$, if $b = c$, then $x_j \cdot x_k = E$, and so the expression for E as k 2-cycles reduces to an expression with $(k-2)$ 2-cycles. This proves the claim in this case. Otherwise, we can suppose that $b \neq c$, and so our expression for E looks like this:

$$E = y_1 \cdot y_2 \cdots y_{k-2} \cdot [ac] \cdot [ab]$$

where the y_i are 2-cycles. The product $[ac] \cdot [ab]$ does not fix a, and so the point a must occur among the $(k-2)$ 2-cycles to the left of $[ac] \cdot [ab]$. Say $y_i = [ad]$. If $d \in \{b, c\}$, then $[ad]$ occurs twice among the k 2-cycles, and this is the same as the case $b = c$ already considered. In that case E is a product of $(k-2)$ 2-cycles.

If $d \notin \{b, c\}$, then move $[ad]$ to the right, preserving the total number of 2-cycles, and obtaining $[ad] \cdot [ac] \cdot [ab]$ at the right end of our expression for E. Since b, c, d are all distinct, the three 2-cycles at the end do not fix a, and so there must be yet another occurrence of the point a in the $(k-3)$ 2-cycles to the left of the three at the end. Continuing this process, there must be an eventual repetition among the points paired with the point a in the 2-cycles moved to the right. In other words, we encounter a 2-cycle $[ae]$ twice in an expression of E as a product of k 2-cycles. As before, we can then write E as a product of $(k-2)$ 2-cycles.

In all cases, we can write E as a product of $(k-2)$ 2-cycles. Induction shows that $(k-2)$ is even, and so k is even. □

The point of writing E as a product of 2-cycles is to define what is called the *parity* of a permutation. Parity can be expressed in several ways: by assigning ± 1 to each permutation, by assigning $0, 1$ to each permutation, or by assigning the words "even, odd" to each permutation. We will use $0, 1$, the elements of $\mathbb{Z}_2$. The Parity Theorem shows how to define this function, and it lists its important properties. We will use the name σ for the parity function.

Many of the statements in the following are redundant; we want to emphasize the properties of the parity function σ. The congruences in the following are mod 2.

PARITY THEOREM. *Let P be a finite set with $|P| \geq 2$. There is a unique function $\sigma : \mathbb{S}_P \to \mathbb{Z}_2$ such that*

(a) If $x \in \mathbb{S}_P$ is the product of an even number of 2-cycles, then $\sigma(x) \equiv 0$.
(b) If $x \in \mathbb{S}_P$ is the product of an odd number of 2-cycles, then $\sigma(x) \equiv 1$.
(c) If $x \in \mathbb{S}_P$ is an r-cycle, then $\sigma(x) \equiv r - 1$.
(d) If $x, y \in \mathbb{S}_P$, then $\sigma(x \cdot y) \equiv \sigma(x) + \sigma(y)$.

PROOF. Let $x \in \mathbb{S}_P$, and Proposition 5.4 shows that x is a product of 2-cycles. We claim that x cannot be a product of an even number of 2-cycles and also a product of an odd number of 2-cycles. For if

$$c_1 \cdots c_j = x = d_1 \cdots d_k$$

where the c_i and d_i are 2-cycles, then observe that

$$x^{-1} = d_k \cdot d_{k-1} \cdots d_1$$

so that
$$E = c_1 \cdots c_j \cdot d_k \cdots d_1$$
and we have written E as a product of $j + k$ 2-cycles. Proposition 5.6 shows that $j + k$ is even, and so either both j, k are even, or both j, k are odd. Thus, if x can be written as the product of an even number of 2-cycles, then all such products have an even number of 2-cycles, and if x can be written as the product of an odd number of 2-cycles, then all such products have an odd number of 2-cycles.

The previous paragraph shows that we can define $\sigma : \mathbb{S}_P \to \mathbb{Z}_2$ by setting $\sigma(x) \equiv 0$ when x is the product of an even number of 2-cycles, and $\sigma(x) \equiv 1$ when x is the product of an odd number of 2-cycles. Thus, (a) and (b) hold.

For (d), let $x, y \in \mathbb{S}_P$ and write each as a product of 2-cycles:
$$x = x_1 \cdots x_j \quad \text{and} \quad y = y_1 \cdots y_k$$
The definition of σ shows that $\sigma(x) \equiv j \mod 2$ and $\sigma(y) \equiv k \mod 2$. Then
$$x \cdot y = x_1 \cdots x_j \cdot y_1 \cdots y_k$$
so that
$$\sigma(x \cdot y) \equiv j + k \equiv \sigma(x) + \sigma(y)$$
as needed to prove (d).

For (c), the proof of Proposition 5.4 showed that an r-cycle can be written as a product of $(r - 1)$ 2-cycles. Statement (c) then follows from (d). □

The equation (d) is very significant; equations like it will occur in several later chapters. Stay tuned!

The permutations $x \in \mathbb{S}_P$ such that $\sigma(x) \equiv 0$ are the *even* permutations. The set of even permutations among $\mathbb{S}_P$ is denoted $\mathbb{A}_P$ and called the *alternating group*. on the set P. When P consists of the positive integers up to n, we will follow the notation $\mathbb{S}_n$ and write $\mathbb{A}_P = \mathbb{A}_n$.

Problems

50. Show that a finite set of order n has exactly $n!$ permutations.[2] (Suggestion. It might be easier to prove a more general statement: if P and Q are finite sets of order n, then there are exactly $n!$ one to one, onto functions $f : P \to Q$.)

51. Define T_n to be the number of 2-cycles in $\mathbb{S}_n$ for each natural number n. Prove that $2 \cdot T_n = n \cdot (n-1)$.

52. Write each of the following permutations in disjoint cycle notation.

(a) $[125][579][153][4][82]$.

(b) The permutation f where

$$\begin{array}{c|ccccccc} x & 1 & 2 & 3 & 4 & 5 & 6 & 7 \\ f(x) & 7 & 6 & 4 & 1 & 2 & 5 & 3 \end{array}$$

53. Let $n \in \mathbb{N}$ and $x \in \mathbb{S}_n$ with $x^2 = E$. Show that x is a product of disjoint 1-cycles and 2-cycles. (Of course, x^2 means "x composed with itself.")

54. Find all $x \in \mathbb{S}_5$ such that $x^2 = [123]$.

55. Let $n \in \mathbb{N}$ and let $x \in \mathbb{S}_n$ be the n-cycle $[123 \cdots n]$. Suppose that $y \in \mathbb{S}_n$ and $x \cdot y = y \cdot x$. Assume that $y(1) = i$. Show that $y(j) \equiv j - 1 + i \bmod n$ for all $1 \leq j \leq n$. (Hint: induction on j.) Note: it follows that there are at most n such y. We will see that there are exactly n such.

56. For $m \in \mathbb{Z}_{12}$, define $f_m : \mathbb{Z}_{12} \to \mathbb{Z}_{12}$ by $f_m(x) \equiv x \cdot m$. Find two values of $m \not\equiv 1$ such that f_m is a permutation. For each of these m's, write f_m as a product of disjoint cycles, and find $k \in \mathbb{Z}_{12}$ such that $f_k = (f_m)^{-1}$.

---

[2]The construction of $n!$ can be done using induction to establish that $1! = 1$, and $(n+1)! = (n+1) \cdot n!$ for all $n \geq 1$.

57. Let $x, y \in \mathbb{S}_n$ be 2-cycles. Show that $x \cdot y$ is E or it is the product of one or two 3-cycles. (Hint: consider whether x, y are the same, have exactly one point in common, or are disjoint. The disjoint case is hard; experiment with products of 3-cycles.)

58. Let $n \geq 3$. Show that every element of $\mathbb{A}_n$ is a product of 3-cycles. (Hint: previous problem.)

59. Let P be a finite set with at least 2 elements, and let $x \in \mathbb{S}_P$. Show that $\sigma(x) = \sigma(x^{-1})$, where σ is the parity function.

60. Let $2 \leq n \in \mathbb{N}$. Show that $|\mathbb{S}_n| = 2 \cdot |\mathbb{A}_n|$.

61. Let A, B be non-empty sets. For $f \in \mathbb{S}_A$ and $g \in \mathbb{S}_B$, define
$$(f \times g) : A \times B \to A \times B \quad \text{by} \quad (f \times g)(a, b) = (f(a), g(b))$$
$$\text{for all} \quad a \in A, \ b \in B$$
Show that $f \times g$ is a permutation on $A \times B$. Let $A = \mathbb{Z}_2 = B$ and show that $A \times B$ has a permutation that is *not* of the form $f \times g$.

62. Find all $x \in \mathbb{S}_4$ such that $x \cdot [1, 2] \cdot [3, 4] = [1, 2] \cdot [3, 4] \cdot x$.

63. Find all $x \in \mathbb{S}_5$ such that $x \cdot [1, 2, 3][4, 5] = [1, 2, 3][4, 5] \cdot x$.

CHAPTER 6

Groups

Recall what it means for a set G to have an operation $\circ$: for all $x, y \in G$, there is $x \circ y \in G$.

DEFINITION 6.1. *Let the set G have an operation $\circ$. Then G is a* group *under $\circ$ if*

(a) $(x \circ y) \circ z = x \circ (y \circ z)$ for all $x, y, z \in G$;
(b) there is $e \in G$ such that $x \circ e = x = e \circ x$ for all $x \in G$;
(c) for every $x \in G$ there is an element $y \in G$ such that $x \circ y = e = y \circ x$.

Statement (a) says that the group operation is associative (as in Chapter 4). An element e as in (b) and (c) is called an *identity element*; a y as in (c) is called an *inverse* for x. We call attention to the quantification in (c). Since x is mentioned before the phrase "there is an element $y \in G$," it is understood that the y referred to depends on x, as is made clear in the equation $x \circ y = e$. Very shortly we will see that such a y is unique (for each x, there exists a *unique y, ...*), we will use the notation x^{-1} for y.

Usually the operation $\circ$ on G is obvious from context, and we will just say, "G is a group," writing xy for $x \circ y$ as we mentioned in Chapter 4. The notation xy for $x \circ y$ suggests an analogy between an abstract group operation and ordinary multiplication. This analogy is somewhat helpful, but we must be sure to stick to the formal definition of a group. For example, the elements $x \circ y$ and $y \circ x$ do not have to be equal! Thus, as in matrix multiplication, you must be careful not to permute group elements multiplied together, although the associative law says that you can bunch them together however you wish.

We first show that the identity element and inverses are unique.

PROPOSITION 6.2. *Let G be a group. Then there is only one identity element of G. Each element of G has a unique inverse. If e is the identity element, and if $xy = e$ for $x, y \in G$, then x and y are the inverses of each other.*

PROOF. Let e and f be identity elements (elements that both satisfy (b) of the definition). That e is an identity forces $ef = f$, that f is an identity implies $ef = e$. Thus $e = ef = f$, so that there is only one identity.

Let $xy = e$. By (c), there is an element z which is an inverse for x. The definition of inverse shows that $zx = e$, so that, using the associative law (a), we get
$$y = ey = (zx)y = z(xy) = ze = z$$
In particular, if y is an inverse for x, then $y = z$ shows that y is the only inverse. But even if we just have $xy = e$, we see from $y = z$ that y is the inverse of x. Since y is the inverse of x, we have $yx = e$ (from (c)). Applying to y the reasoning we used for x, we conclude that x is the inverse of y. □

It is customary to use 1_G or just 1 to denote the identity element of a group G. In doing this, we are using the symbol 1 not as a number, but as a mnemonic device for remembering (b). Given $x \in G$, one uses x^{-1} to stand for the inverse of x. Again, we are employing a mnemonic device; for Proposition 6.2 shows that $(x^{-1})^{-1} = x$, an equation which you have used in number calculations. We do not have fractions, and so x^{-1} should never be written $1/x$.

We emphasize that group elements are not necessarily numbers, and in general you cannot take valid equations concerning numbers and apply them to group elements. For example, if x and y are numbers, then $(xy)^{-1} = (x^{-1})(y^{-1})$ For group elements, the correct equation is $(xy)^{-1} = (y^{-1})(x^{-1})$, as you should verify. Since $(y^{-1})(x^{-1})$ is in general different from $(x^{-1})(y^{-1})$, the number equation is not correct for groups.

Now we find some examples of groups. They originally arose as sets of permutations; they were recognized in a variety of settings by Lagrange, Cauchy, and others during the end of the eighteenth century and the beginning of the nineteenth. Chapter 5 introduced you to the set $\mathbb{S}_P$.

PROPOSITION 6.3. *Let P be a non-empty finite set. Then $\mathbb{S}_P$ is a group under function composition.*

PROOF. Let $f, g \in \mathbb{S}_P$. Since f and g are one to one and onto, Proposition 4.2 shows that fg is one to one and onto. Thus, function composition is an operation on $\mathbb{S}_P$. Proposition 4.1 says that function composition is associative. The identity permutation E_P satisfies $fE_P = f = E_P f$ for every $f \in \mathbb{S}_P$, as pointed out in Chapter 4, and so E_P is an identity element for $\mathbb{S}_P$. If $f \in \mathbb{S}_P$, then f^{-1} is one to one and onto, by Proposition 4.3, and so $f^{-1} \in \mathbb{S}_P$. $\square$

It is a good exercise for you to cover the same ground with $\mathbb{A}_P$. Use the parity function σ to show that function composition is an operation on $\mathbb{A}_P$. Then you can show that $\mathbb{A}_P$ is a group under function composition.

We want to introduce groups of *symmetries*. Symmetries arise naturally in geometric shapes: squares, triangles, circles, etc. Although geometry is important and beautiful, we do not wish to digress to introduce it formally, and so we will study symmetries of *graphs* – not graphs of functions as in Chapter 4, but finite graphs as in combinatorics and computer science. On the next page is a picture of an example; we will refer to it as a *square*, for obvious reasons.

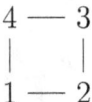

The numbers $1, 2, 3, 4$ are called *vertices*; they are being used merely as labels. The connecting segments are *edges*. Formally, an edge is a subset of order 2, so that the edge connecting 1 and 2 is the set $\{1, 2\}$. In some graphs, edges have a *direction* to them. If the direction goes from 1 to 2, the edge could be the ordered pair $(1, 2)$, and it could be repsented by notation such as $1 \to 2$. To keep things simple, we will work with *undirected* graphs, in which the edges do not have a direction.

In general, if V is a non-empty finite set, then a *graph on V* regards V as the set of vertices and consists of a collection of edges: subsets of V of order 2. For $a, b \in V$, we write $a \leftrightarrow b$ when $\{a, b\}$ is an edge. Notice that $a \leftrightarrow b$ if and only if $b \leftrightarrow a$. Also, $a \leftrightarrow b$ implies that $a \neq b$.

A *symmetry* of the graph V is a permutation f of V such that if $a \leftrightarrow b$ is an edge of V, then $f(a) \leftrightarrow f(b)$ is also an edge. We say that f *preserves edges*. The proof of the following is a simple matter of following definitions closely. Once we get through the proof, we can give many examples.

PROPOSITION 6.4. *A symmetry of a graph maps the set of edges of the graph one to one, onto itself. The set of symmetries of a graph forms a group under function composition.*

PROOF. Let G be the set of symmetries of a graph V. Let $f \in G$, and define a mapping F from the set of edges of the graph to itself by $F(a \leftrightarrow b) = f(a) \leftrightarrow f(b)$. Because f is one to one, it is easy to see that F is one to one. Since the set of edges is finite, Proposition 5.1 shows that F is a permutation of the set of edges.

6. GROUPS

Next we claim that if $f \in G$, then $f^{-1} \in G$. Indeed, use the mapping F defined in the previous paragraph. If $a \leftrightarrow b$ is in the graph, since F is onto there is $c \leftrightarrow d$ in the graph such that $F(c \leftrightarrow d) = a \leftrightarrow b$. In other words, $f(c) \leftrightarrow f(d) = a \leftrightarrow b$. This implies that either $f(c) = a$ and $f(d) = b$, or that $f(c) = b$ and $f(d) = a$. Assume, without loss of generality, that $f(c) = a$ and $f(d) = b$. Then $c = f^{-1}(a)$ and $d = f^{-1}(b)$. The fact that $c \leftrightarrow d$ is in the graph is then that $f^{-1}(a) \leftrightarrow f^{-1}(b)$ is in the graph. We have proved that if $a \leftrightarrow b$ is in the graph, then so is $f^{-1}(a) \leftrightarrow f^{-1}(b)$. Thus, $f^{-1} \in G$.

We need to show that function composition is an operation on G. Let $f, g \in G$, and then fg is a permutation of V. Let $a \leftrightarrow b$ be an edge in the graph. Then $g(a) \leftrightarrow g(b)$ is also in the graph, and then $f(g(a)) \leftrightarrow f(g(b))$ is in the graph as well. In other words, if $a \leftrightarrow b$ is in the graph, then so is $fg(a) \leftrightarrow fg(b)$. Therefore $fg \in G$. The operation on G is associative, because function composition is always associative (Proposition 4.1).

The identity permutation E_V on V preserves edges, and so $E_V \in G$, and G has an identity element. We have already shown that the inverses of elements of G are in G. □

Let V be the square defined above. We let D_8 be the group of symmetries (the 8 will be explained momentarily). Let's find the elements of D_8. Notice that the permutation

$$R = [1, 2, 3, 4]$$

is a symmetry. The graph has four edges; make sure you see that R maps each of these edges to another edge! We used R here for *rotation*, since R "rotates" the square 90 degrees counterclockwise, so R has a geometric interpretation. Continuing to rotate the square, or, alternatively, composing R with itself we obtain three other symmetries: R^2, R^3, and R^4. Notice that $R^4 = E_4$, the identity function.

You can also get symmetries by "flipping the square over about an axis." For instance, define

$$F = [1,4][2,3]$$

and observe that $F \in D_8$. The symmetry F flips the square about the "axis" perpendicular to the edges $1 \leftrightarrow 4$ and $2 \leftrightarrow 3$. Composing F with the rotations gives us four more symmetries, each of which is called a *reflection*: F, FR, FR^2, and FR^3. See if you can find an axis for each of the reflections. We now have 8 different elements of D_8: the four rotations and the four reflections.

The significance of the 8 in D_8 is that there are exactly 8 symmetries. Proving this illustrates an important principle of counting. Given $f \in D_8$, there are at most four vertices that can be $f(1)$. Once $f(1)$ is chosen, the definition of symmetry shows that $f(1) \leftrightarrow f(2)$ must be in the graph. There are two edges involving $f(1)$, and so there are a maximum of two possibilities for $f(2)$, given $f(1)$. We claim that the placement of $f(1)$ and $f(2)$ determines the placement of the remaining two corners $f(3)$ and $f(4)$. Indeed, $f(2) \leftrightarrow f(3)$ must be an edge. Since $f(1) \leftrightarrow f(2)$ is already an edge, and since $f(3)$ cannot equal $f(1)$ (since f has to be one to one), we see that $f(3)$ has to be the unique vertex not equal to $f(1)$ in an edge with $f(2)$. Then $f(4)$ is the remaining vertex.

Thus, a symmetry has 4 choices for $f(1)$ and then, given $f(1)$, it has at most 2 choices for $f(2)$. The vertices $f(1)$ and $f(2)$ determine the other vertices $f(3)$ and $f(4)$, so that there are no more choices. This proves that there are at most $4 \cdot 2 = 8$ elements in D_8. Since we have already produced 8 distinct elements in D_8 (so that $|D_8| \geq 8$), we see that $|D_8| = 8$.

Let Q be the following graph, called a triangle.

```
    3
   / \
  1 — 2
```

The symmetry group is called D_6. You will show that D_6 consists of three rotations and three reflections, so that $|D_6| = 6$.

Because D_8 and D_6 are associated with graphs that mimic plane figures, they are called *dihedral groups*.[1] The word *dihedral* explains the capital D used in the names of these groups.

Another example of a group is the integers under the operation addition. Notice that the Integer Axioms 2.1(1,2,3) say, word for word, that $\mathbb{Z}$ is a group with addition as operation. The identity element is 0, so that the customary use of "1" for the identity is confusing. The fact that the integers is a group under addition does not give us any real insight; for one thing, it ignores multiplication. The concept of a *ring*, introduced in Chapter 14, will better generalize the integers.

A slightly more interesting example: for each positive integer n, the set $\mathbb{Z}_n$ is a group under addition. This is the content of Proposition 3.2. As with the integers, the identity element 0 should not be called "1" and we had better stick to $x + y$ rather than xy to denote the operation. As with the integers, it will be better to understand $\mathbb{Z}_n$ as a ring.

Here is an important example using multiplication in $\mathbb{Z}_n$. Let n be a positive integer modulus, and define U_n to be the set of all $x \in \mathbb{Z}_n$ such that x has a multiplicative inverse in $\mathbb{Z}_n$. This is the *units group mod n*; we claim it is a group under multiplication in $\mathbb{Z}_n$. If $x, y \in U_n$, then $x \cdot y$ has inverse $y^{-1} \cdot x^{-1}$, and so $x \cdot y \in U_n$. Thus, multiplication is an operation on U_n. Next, since $1 \cdot 1 \equiv 1$, we see that $1 \in U_n$, and obviously 1 is the multiplicative identity. Finally, if $x \in U_n$, then $x^{-1} \in U_n$, as well, since the inverse of x^{-1} is x. This proves that U_n is a group under multiplication.

Compute that

$$U_2 = \{1\}, \quad U_3 = \{1, 2\}, \quad U_8 = \{1, 3, 5, 7\}$$

---

[1] The word *dihedral* means *two-sided*.

For $x \in U_n$, there is a unique inverse $x^{-1} \in U_n$. In this notation, x^{-1} depends on n. We have that $2 \in U_5$ with $2^{-1} = 3$. Also $2 \in U_7$, but now $2^{-1} = 4 \neq 3$. Also $2 \notin U_8$, since 2 has no inverse mod 8. The inverse notation needs to be used with care.

The group G is *abelian* if $xy = yx$ for all $x, y \in G$. Which of the groups that we have defined so far are abelian?

Here is a way to build new groups from old ones. If G and H are groups, we can make the Cartesian product $G \times H$ into a group by setting

$$(g_1, h_1) \circ (g_2, h_2) = (g_1 g_2, h_1 h_2) \quad \text{for all} \quad g_1, g_2 \in G, \ h_1, h_2 \in H$$

Notice that $g_1 g_2$ is computed using the operation of G, and $h_1 h_2$ using that of H. It is easy to check that the definition of group is satisfied. For example, the identity element is $(1_G, 1_H)$. Under these circumstances, we call $G \times H$ the *direct product* G and H.

We close with a notational matter. Given an element x of a group G, we write $x^1 = x$, $x^2 = xx$, $x^3 = xxx$, and so on. This notation makes sense because the associative law tells us it doesn't matter in what order the product is actually computed. Thus, we can write x^n for $n \in \mathbb{N}$. We will also write $x^0 = 1$, but we caution you that we are not fully introducing exponential notation; we will discuss negative exponents later, and, as we have already hinted, expressions such as $(xy)^n$ are problematic. A permutation f is already a function from a set to itself, and we have previously introduced the notation f^n in that case. (Compose f over and over, n times.)

One of the best books on group theory is *Theory of Groups of Finite Order*, by W. Burnside (Dover, 1955). It is one of those books that is difficult but teaches real insight.

Problems

64. Let G be a group and $x, y \in G$. Show that $(x \cdot y)^{-1} = y^{-1} \cdot x^{-1}$.

65. Let G be a group. Show that G is abelian if and only if $(x \cdot y)^2 = x^2 \cdot y^2$ for all $x, y \in G$.

66. Suppose that G is a group and $x^2 = 1_G$ for all $x \in G$. Show that G is abelian.

67. Let sets X, Y, define $X \setminus Y$ to be the set of elements of X that are not in Y. (So, $x \in X \setminus Y$ if and only if $x \in X$ and $x \notin Y$.) Let A be a set, and let G be the set of all its subsets.[2] Define an operation on G by the formula

$$X * Y = (X \setminus Y) \cup (Y \setminus X) \quad \text{for all} \quad X, Y \in G$$

Show that G is a group under this operation, and that $X^2 = 1_G$ for all $X \in G$.

68. Let f be a symmetry of the finite graph V, and let x be a vertex. Show that x and $f(x)$ are involved in the same number of edges.

69. Find the elements of D_6, showing that there are exactly 6 of them.

70. Find the symmetries of the graph pictured below. This group is called the *Klein Four-Group*, denoted K_4. (Note: $3, 4$ are vertices not connected to any edges.) You should also show that $K_4 \subseteq D_8$.

```
4    3

1 ── 2
```

71. Working by analogy with D_8 and D_6, invent a graph that looks like a pentagon and has exactly 10 symmetries. (Its symmetry group is called D_{10}.)

72. Show that the even integers form a group under addition, carefully noting which integer axioms are being used.

73. Let G be a group and let $x, y \in G$. Show that $(y \cdot x \cdot y^{-1})^k = y \cdot x^k \cdot y^{-1}$, for every non-negative integer k. (Hint: induction on k.)

---

[2] It is a matter of set theory that for each set there is a set of all its subsets. When the set is finite of order n, we already know that it has exactly 2^n subsets.

74. Let G be a group. Define $\sigma : G \to G$ by the formula $\sigma(x) = x^{-1}$. Show that σ is a permutation. Now assume that G is finite and $|G|$ is even. Show that σ has at least two fixed points. Conclude that there is $x \in G$ with $x \neq 1$ and $x^2 = 1$. (Hint: what is the cycle structure of σ?)

75. Let G be a group, and let $y \in G$. Define $f : G \to G$ by the formula $f(x) = x \cdot y$ for all $x \in G$. Show that f is a permutation. Find the cycles of this permutation when $G = \mathbb{Z}_{12}$ and $y = 10$. (Remember that the operation on $\mathbb{Z}_{12}$ is addition.)

76. Show that the direct product $G \times H$ of groups G, H is a group, as claimed on p.64.

77. Show that $\mathbb{A}_4$ has three elements that can be written as 2 disjoint 2-cycles, and show that these elements commute with each other.

78. Let n be a positive integer, and let m be an integer. Let G be the set of all $x \in \mathbb{Z}_n$ such that $m \cdot x \equiv 0$. Show that G is a group under addition. (Note: start by showing that addition is an *operation* on G.) Find G explicitly when $n = 24$ and $m = 20$.

CHAPTER 7

A Single Group Element

Throughout this chapter x will be a specific element of a specific group G. We have introduced the exponential notation x^n, when n is a non-negative integer. As we have remarked, the associative law of group multiplication shows that it does not matter how the n copies of x are grouped together to perform the multiplication. Given non-negative integers n and m, multiplying n copies of x by m copies of x is the same as multiplying $n + m$ copies of x. Thus

$$(x^n)(x^m) = x^{n+m}$$

Furthermore, if $y = x^n$, then multiplying m copies of y is the same as multiplying nm copies of x, so that

$$(x^n)^m = x^{nm}$$

These equations are the usual rules of exponents to a common base. We wish to expand them to hold for negative integers as well. This is not really hard, but it does take some care (and a proof).

Note the following definition carefully: for a negative integer n, we define

$$x^n = (x^{-1})^{-n}$$

which is $-n$ copies of x^{-1}. Notice that x^{-1} can now be read in two ways, as "the inverse of x" and as "x to the minus one power." Since the definition of the latter is "x inverse to the first power" the two formulations agree.

PROPOSITION 7.1. *Let G be a group, $x \in G$, and let m and n be integers. Then*

a) $(x^n)(x^m) = x^{n+m}$;

b) $(x^n)^m = x^{nm}$.

PROOF. For (a), let us first do the case where both m and n are negative. Then by definition
$$(x^n)(x^m) = (x^{-1})^{-n}(x^{-1})^{-m}$$
Observe that $-n$ and $-m$ are positive, so that
$$(x^{-1})^{-n}(x^{-1})^{-m} = (x^{-1})^{-n+-m} = (x^{-1})^{-(n+m)}$$
Since $m+n < 0$, it is a definition that
$$(x^{-1})^{-(n+m)} = x^{n+m}$$
which produces (a).

If either m or n is 0, then the statement (a) is trivial. It remains to assume that one of m and n is positive while the other is negative. Without loss of generality, let $n > 0 > m$. Then $x^m = (x^{-1})^{-m}$, noting that $-m > 0$. If $n \geq -m$, then $(x^n)(x^m) = x^n(x^{-1})^{-m}$ means canceling $-m$ copies of x from x^n, the result would be $n - (-m) = n + m$ copies of x, in line with (a). If $n < -m$, then
$$(x^n)(x^m) = x^n(x^{-1})^{-m} = x^n(x^{-1})^n(x^{-1})^{-m-n}$$
(notice that $-m - n > 0$) means canceling all the n copies of x from x^n and then having $(-m) - n$ copies of x^{-1} left over. In this case
$$(x^n)(x^m) = (x^{-1})^{-n-m} = x^{-(-n-m)} = x^{n+m}$$
Statement (a) is proved.

A similar argument will prove (b). You might try it yourself. □

One consequence of Proposition 7.1 is that $x^n \cdot x^{-n} = x^0 = 1$. This tells us that the inverse of x^n is x^{-n}. This convenient fact accords with number algebra.

The object of this chapter is to understand the set of all powers of a particular group element. You must keep track of two things: the definition of this set, and then an important simplification that we will discover after working with the set. Here is the definition:

$$\langle x \rangle = \{x^n \mid n \in \mathbb{Z}\}$$

The angle brackets $\langle$ and $\rangle$ are delimiters and they have nothing to do with "less than" or "greater than." Proposition 7.1 shows for x^n and x^m in $\langle x \rangle$, that their product $(x^n)(x^m) = x^{n+m}$ is also in $\langle x \rangle$. Thus the group multiplication is an operation on $\langle x \rangle$. Furthermore, $\langle x \rangle$ has an identity element, namely $1 = x^0$, and given $x^n \in \langle x \rangle$, its inverse x^{-n} is also in $\langle x \rangle$. We see that $\langle x \rangle$ is itself a group; we say that it is the *cyclic group generated by x*.

A group H is *cyclic* if there is $x \in H$ such that $H = \langle x \rangle$, and we say that x is a *generator* of H. There are several facts we wish to cover in this setting; for now we point out that the generator of a cyclic group *is not, in general, unique*. For instance, $\langle [12345] \rangle$ has four generators; it is a nice exercise in understanding the definitions given so far to find them! Secondly, the construction of $\langle x \rangle$ began with a group G where $x \in G$. It is usually the case that $\langle x \rangle$ is relatively small compared to G, so that G has lots of elements not included in $\langle x \rangle$. We chose $[12345]$ from $\mathbb{S}_5$; there are 115 elements of $\mathbb{S}_5$ that are not in $\langle [12345] \rangle$.

Since Proposition 7.1 shows that the powers of a group element x commute, we see that cyclic groups are always abelian. Thus, for instance, D_8 and D_6 are not cyclic. You should show that although K_4 is abelian, it is not cyclic. Cyclic groups are special.

To take another example, consider the element $x = [123][45]$ of $\mathbb{S}_5$. We can compute the positive and negative powers of this element:

$$x = [123][45] \qquad x^2 = [132] \qquad x^3 = [45]$$
$$x^4 = [123] \qquad x^5 = [132][45] \qquad x^6 = E$$
$$x^7 = x^6 \cdot x = x \qquad x^8 = x^6 \cdot x^2 = x^2 \qquad \ldots$$
$$x^{-1} = [132][45] = x^5 \qquad x^{-2} = (x^5)^2 = x^4 \qquad x^{-3} = (x^5)^3 = x^3$$
$$\ldots$$

We see the powers x, x^2, x^3, x^4, x^5, E just keep repeating over and over. Thus, the set of *all* powers of x is really just a set with 6 elements:

$$\langle [123][45] \rangle = \{E, [123][45], [132], [45], [123], [132][45]\}$$

Here is another example: the element 5 in $\mathbb{Z}_{20}$. Remembering that the operation is addition, the "powers" of 5 are 5 and $5 + 5$ and $5 + 5 + 5$ and so on. The inverse of 5 is $-5 \equiv 15$. Here are the positive powers on the top line and the negative powers on the bottom line.

| 5 | 10 | 15 | $20 \equiv 0$ | $25 \equiv 5$ | $30 \equiv 10$ | $\ldots$ |
| 15 | $30 \equiv 10$ | $45 \equiv 5$ | $60 \equiv 0$ | $75 \equiv 15$ | $90 \equiv 10$ | $\ldots$ |

And we see that

$$\langle 5 \rangle = \{0, 5, 10, 15\} \quad \text{in} \quad \mathbb{Z}_{20}$$

In each case so far considered the set $\langle x \rangle$ has been finite. This is not always the case, but since we have already given a strong hint for that case, we pursue it explicitly. Assume, then, that $\langle x \rangle$ is finite. For instance, if G is a finite group and $x \in G$, then since $\langle x \rangle$ is a subset of G, the cyclic group would be finite. If we study $\langle x \rangle$ we will see a connection with the groups $\mathbb{Z}_n$ and with the cycles of a permutation. Because $\langle x \rangle$ is finite, the list

$$1, \quad x, \quad x^2, \quad x^3, \quad \ldots$$

must repeat at some point. Thus there are non-negative integers $p < q$ such that $x^p = x^q$. Let q be minimal with this property and we claim that $x^q = 1$ (so that $p = 0$). Indeed, if $p \geq 1$, then multiplying $x^p = x^q$ by x^{-1} gives $x^{p-1} = x^{q-1}$. Since $p \geq 1$, the number $p - 1$ is greater than or equal to 0, hence x^{q-1} is an earlier repeat than x^q, a contradiction to the minimality of q. Thus $p = 0$, so that $x^q = 1$. The number q is greater than $p = 0$, and so q is a positive integer. We see that q is the smallest positive integer for which $x^q = 1$. This number q is called the *order* of x, and denoted $o(x)$.

The order of $[123][45]$ is 6; the order of $5 \in \mathbb{Z}_{20}$ is 4.

Here is a complete description of $\langle x \rangle$.

PROPOSITION 7.2. *Let x be an element of a group G, and suppose that $\langle x \rangle$ is finite. For each integer m there is $r \in \mathbb{Z}$ with $0 \leq r < o(x)$ and $x^m = x^r$. We have*
$$\langle x \rangle = \{1,\ x,\ x^2,\ \ldots,\ x^{o(x)-1}\}$$
and the elements listed in this set are distinct, so that $|\langle x \rangle| = o(x)$.

PROOF. Given the integer m, use the Division Theorem to write $m = o(x) \cdot q + r$ where $0 \leq r < o(x)$. Compute
$$x^m = x^{o(x) \cdot q + r} = (x^{o(x) \cdot q}) \cdot (x^r)$$
$$= (x^{o(x)})^q \cdot (x^r) = 1^q \cdot (x^r) = x^r$$

This proves that $\langle x \rangle$, defined as the set of x^m for all integers m, is really only the set of x^r where $0 \leq r < o(x)$, which is the set listed in the conclusion of the Proposition.

We must show that $|\langle x \rangle| = o(x)$; in other words that the x^r are distinct for distinct r with $0 \leq r < o(x)$. But we have already done this (!), for when we listed $1, x, x^2, \ldots$, the element $x^{o(x)}$ was the first repetition in this list. Thus the x^r are distinct for distinct r, so that $\langle x \rangle$ has exactly $o(x)$ elements. $\square$

Continuing the notation of Proposition 7.2, use $o(x)$ as modulus. You will show for integers a and b, that $a \equiv b$ if and only if $x^a = x^b$. This yields a one to one, onto function from $\mathbb{Z}_n$ to $\langle x \rangle$, such that a corresponds to x^a. Think about this.

Note that we are using the word *order* in two different ways; we speak of the "order of a set" and the "order of a group element." This double usage is unfortunately standard; at least the equation $o(x) = |\langle x \rangle|$ shows that the uses are related.

We mention one further result.

COROLLARY 7.3. *Let x be an element of a group, and assume that $\langle x \rangle$ is finite. Then $x^m = 1$ if and only if $o(x)$ divides m.*

Corollary 7.3 underscores the fact that an equation $x^m = 1$ does **not** force $m = o(x)$, **only** that $o(x)$ divides m.

It will be helpful to consider the case of the group $\mathbb{Z}_n$, in which we do not use the exponential notation. For $x \in \mathbb{Z}_n$ the group notation x^m would mean $x + x + \cdots + x$ where there are m terms. We might wish to write this as mx. What is $o(x)$ for such an x? Experiment with a few cases to find some sort of pattern. We will give a complete answer later.

The order of a permutation can be deduced easily from its cycles. You probably know what is meant by the *least common multiple* of a non-empty, finite set S of natural numbers. If S' be the set of all natural numbers divisible by all the members of S, then the least common multiple of S is the minimum of S'. This minimum exists by well-ordering, since the set S' is non-empty, for it contains the product of all the elements of S. As an exercise from the next chapter, you will prove that if all the elements of S divide a natural number n, then n is divisible by the least common multiple of S.

PROPOSITION 7.4. *The order of a permutation that is an n-cycle is n. The order of a permutation is the least common multiple of the set of lengths of its disjoint cycles.*

PROOF. We leave the case of an n-cycle to you.

Write the permutation f as a product of disjoint cycles, as in Proposition 5.3:

$$f = c_1 \cdot c_2 \cdots c_m$$

and let the length of c_i be n_i. Let N be the least common multiple of the n_i, and let n be the order of f. We will show that $n = N$.

We have $f^n = E$ (the identity function). Proposition 5.2 shows that the distinct c_i commute, and so we have

$$E = f^n = c_1^n \cdots c_m^n$$

Each c_i^n must be E, since if some c_i^n does not fix the point p, then because the cycles are disjoint, we would have that f^n does not fix p, and then $f^n \neq E$. Thus, each $c_i^n = E$. You have shown that the order of c_i is its length n_i, and so Corollary 7.3 shows that n_i divides n. Therefore, since N is the least common multiple of the n_i, it divides n.

Conversely, observe that

$$f^N = c_1^N \cdots c_m^N$$

Since n_i divides N for each i, we have $c_i^N = E$, by Corollary 7.3. Thus, $f^N = E$, and so, Corollary 7.3 again!, n divides N.

The two natural numbers n and N each divide the other, and so $n = N$. □

We have been assuming that $\langle x \rangle$ is finite. What if it isn't? The best we can do in this case is to stick to the original definition of $\langle x \rangle$ as the set of *all* integer powers of x. In this case, we define the order of x to be "infinity," and we write $o(x) = \infty$. The only example of an infinite group we have so far is

$\mathbb{Z}$, and in that group 1 has order infinity. We will stick almost exclusively to finite groups in this course.

A group element of order 2 is called an *involution*. The involutions in a group are important in many ways. Here is a general way to build a group out of 2 involutions. Let G be a finite group, and let $x, y \in G$ be distinct, each having order 2. Notice that

$$(x \cdot y) \cdot (y \cdot x) = x \cdot (y \cdot y) \cdot x = x \cdot 1 \cdot x = x \cdot x = 1$$

so that xy and yx are inverses. Let $z = xy$. Because $x \neq y$, we see that $z \neq 1$. Also $z^{-1} = yx$. Since G is finite, the element z has finite order, say $n = o(z)$, and so $n > 1$. Let H be the set of group elements $z^a \cdot x^b$ where $a \in \mathbb{Z}_n$ and $b \in \mathbb{Z}_2$. We claim that H is a group, using the group operation of G, and that $|H| = 2 \cdot n$.

First, notice that $x \cdot z = x \cdot x \cdot y = y = y \cdot x \cdot x = z^{-1} \cdot x$. It follows by induction that if $a \in \mathbb{Z}$, then $x \cdot z^a = z^{-a} \cdot x$. Now we can show that the group operation is an operation on H, for if $z^a \cdot x^b$ and $z^c \cdot x^d$ are given, then their product is in H. Indeed, if $b \equiv 0 \mod 2$, then

$$z^a \cdot x^b \cdot z^c \cdot x^d = z^a \cdot z^c \cdot x^d = z^{a+c} \cdot x^d$$

and if $b \equiv 1 \mod 2$, then

$$z^a \cdot x^b \cdot z^c \cdot x^d = z^a \cdot x \cdot z^c \cdot x^d = z^a \cdot z^{-c} \cdot x \cdot x^d = z^{a-c} \cdot x^{1+d}$$

The associative law holds in H because it holds in G. The identity element is $1 = z^0 \cdot x^0$. The inverse of $z^a \cdot x^b$ is z^{-a} if $b \equiv 0 \mod 2$, and it is

$$x \cdot z^{-a} = z^a \cdot x \quad \text{if} \quad b \equiv 1 \mod 2$$

Thus, H is a group. As to the order of H, we claim that $x \notin \langle z \rangle$. For if $x \in \langle z \rangle$, then x, z commute. But $x \cdot z = z^{-1} \cdot x$, and so $z = z^{-1}$, whence z has order 2. The element x has order 2 in $\langle z \rangle$, and so $x = z$. Since $z = x \cdot y$, this

implies that $y = 1$, a contradiction. We conclude that $x \notin \langle z \rangle$. Now suppose that $z^a \cdot x^b = z^c \cdot x^d$, so that
$$z^{a-c} = x^{d-b}$$
Since $x \notin \langle z \rangle$, we see that $d - b \equiv 0 \mod 2$, and so $x^b = x^d$. We also have $z^{a-c} = 1$, and Corollary 7.3 says that $a - c \equiv 0 \mod n$, so that $z^a = z^c$. This shows that as a runs over $\mathbb{Z}_n$ and b runs over $\mathbb{Z}_2$, the elements $z^a \cdot x^b$ of H are distinct. Thus, $|H| = 2 \cdot n$.

The group H is denoted $\langle x, y \rangle$, because it comes about from the elements x, y.

We have one more group to introduce: the *quaternion group* Q_8. We will give an abstract construction and then realize that construction in permutations. We suppose we have a group G with an element x of order 4, and we write $x^2 = z$, so that z has order 2. Assume there is another element $y \in G$ with $y^2 = z$. We claim that y has order 4. Indeed, $y^4 = (y^2)^2 = z^2 = x^4 = 1$. This shows that the order of y *divides* 4. Since $z \neq 1$, we have $y^2 \neq 1$, and so y cannot have order 1 or 2. Thus, y has order 4. Next, we assume that $(xy)^2 = z$, and then xy has order 4, as well. Define
$$Q_8 = \{x^a \cdot y^b \mid 0 \leq a \leq 3,\ 0 \leq b \leq 1\}$$
We claim that Q_8 is a group of order 8. This will be taken up in class or on homework. As a first calculation, you should show that $y \cdot x = x \cdot y \cdot z$. (Hint: multiply the left side of both sides of the equation by xy.)

Problems

79. Prove Proposition 7.1b. (Hint: for $m \geq 0$ and n arbitrary, use induction and 7.1a. If $m < 0$, then $m = -(-m)$.)

80. Find the order of every element of the cyclic group $\langle [123456] \rangle$, and indicate the generators of that group.

7. A SINGLE GROUP ELEMENT

81. Find the order of every element of $\mathbb{Z}_{14}$.

82. Find the order of every element of the direct product $\mathbb{Z}_2 \times \mathbb{Z}_4$.

83. What is the order of 3 as an element of the integers under addition? As an element of $\mathbb{Z}_7$ as a group under addition?

84. Use the cycle structures to write down the list of orders of elements of $\mathbb{S}_4$. (You don't need to write down all the elements, just all the possible orders.)

85. What is the smallest positive integer n such that $\mathbb{S}_n$ has an element of order 6?

86. Let G be a finite group, and let $x, y \in G$. Show that the order of x is the same as the order of $y \cdot x \cdot y^{-1}$. (Recall from a problem in Chapter 6 that we know how to compute powers of $y \cdot x \cdot y^{-1}$.)

87. Show that the order of each element of D_8 divides the order of that group (as a set). Same problem for $\mathbb{S}_3$.

88. The notation $\langle [12][34], [13] \rangle$ indicates the smallest group containing both $[12][34]$ and $[13]$. Show that $\langle [12][34], [13] \rangle = D_8$.

89. Show that $\langle [12], [13] \rangle = \mathbb{S}_3$.

90. Find all the elements of $\langle [12][34], [13][56] \rangle$.

91. Find $x, y \in \mathbb{S}_5$ such that x is a 3-cycle, y is a 3-cycle, $x \cdot y$ is a 5-cycle.

92. Try to build a group out of $[123]$ and $[34]$. How many elements do you get?

93. Let $x = [1,2,3,4][5,6,7,8]$ and $y = [1,7,3,5][2,6,4,8]$. Show that $x^2 = y^2 = (xy)^2$. Write down the elements of Q_8 in this case.

94. Let $x \in G$ where G is a group. Show that x and x^2 have the same order if and only if the order of x is odd.

CHAPTER 8

Factorization

The *greatest common divisor* (*GCD*) of given integers n and m is the largest natural number that divides both of them. The existence of such a number is expressed in the following, along with other facts. The most important part of this result is the mysterious statement (b).

GCD THEOREM. *Let m and n be integers, not both 0. Then*
(a) m and n have a GCD;
(b) the GCD of m and n can be written in the form $c \cdot m + d \cdot n$ for some integers c and d.
(c) If the integer b divides both m and n, then b divides the GCD.

PROOF. As in the proof of the Division Theorem, the idea is a clever use of induction, which, for this theorem, goes all the way back to Euclid.[1] Let
$$S = \left\{ y \in \mathbb{N} \mid y = c \cdot m + d \cdot n \text{ for some } c, d \in \mathbb{Z} \right\}$$
We show that S is not empty. Indeed, if m is not zero, then $1 \cdot m + 0 \cdot n = m$ and $-1 \cdot m + 0 \cdot n = -m$, and one of these is positive, hence in S. Similarly, S is not empty if n is not zero. Anyway, well-ordering gives to S a minimal element a, and we will show that a is the unique GCD. Observe that $a = cm + dn$ for some $c, d \in \mathbb{Z}$, by the definition of S.

Does a divide n? The Division Theorem grants $n = q \cdot a + r$ where $0 \leq r < a$. Suppose that r is not zero (that a does not divide n), and use $a = cm + dn$ to

---

[1]See the bibliographic reference at the end of this chapter.

write $n = q \cdot (cm + dn) + r$, which is $-q \cdot c \cdot m + (1 - q \cdot d) \cdot n = r$. We have that $r > 0$ and r is an integer combination of n and m, thus $r \in S$. But $r < a$ violates the minimality of a. This contradiction shows that $r = 0$, whence $n = q \cdot a$ is divisible by a.

Similarly, a divides m.

Let b divide both n and m, writing $n = p \cdot b$ and $m = q \cdot b$. Then

$$a = c \cdot m + d \cdot n = c \cdot q \cdot b + d \cdot p \cdot b = (c \cdot q + d \cdot p) \cdot b$$

shows that b divides a, which is statement (c). In particular, since $a > 0$, we have $b \leq a$, and this shows that a is the GCD. $\square$

Observe that 0 has no greatest common divisor with itself, since every integer divides 0, so we need the hypothesis that one of m, n is not zero.

If $n \neq 0$, then the natural number divisors of n are bounded by $|n|$ (absolute value). Thus, the common divisors of n and m are bounded above. And since 1 is a common divisor, the set of common divisors is not empty. It follows from well-ordering that there is a maximal (greatest!) common divisor. That's a shorter proof, but it turns out that statement (b) is crucial to our use of the GCD, and so we have used Euclid's proof with its emphasis on the $c \cdot m + d \cdot n$ formula.

Here is an important special case. We will leave the proof to you or to class.

PROPOSITION 8.1. *Let a, b be integers, not both 0. Then a, b have GCD equal to 1 if and only if there are integers c, d such that $a \cdot c + b \cdot d = 1$.*

We turn to an application of the GCD to group theory. For a positive integer n, we will need to see that if $a \equiv b$ mod n, then a and b each has the same GCD with n. We leave this as an exercise.

8. FACTORIZATION

PROPOSITION 8.2. *Let $n \in \mathbb{N}$ and $x \in \mathbb{Z}$. Then $x \in U_n$ if and only if n and x have GCD equal to 1.*

PROOF. We will prove one direction: let n and x have GCD equal to 1, and find integers a, b such that $a \cdot n + b \cdot x = 1$. Reading this in $\mathbb{Z}_n$, we have $a \cdot 0 + b \cdot x \equiv 1$, or $b \cdot x \equiv 1$. Thus, x has b as multiplicative inverse, so that $x \in U_n$.

The other direction of the proof will be done in class or as homework. □

As a consequence of what we have done, we find the order of a power of a group element. This result is tricky; apply it very carefully.

PROPOSITION 8.3. *Let G be a group, let $x \in G$ and let $o(x)$ be finite. Let m be an integer, let a be the GCD of $o(x)$ and m, and write $o(x) = a \cdot b$ for an integer b. Then $o(x^m) = b$.*

PROOF. The GCD of $o(x)$ and m exists, since $o(x)$ is not zero. Furthermore that a divides $o(x)$ produces the equation $o(x) = a \cdot b$. Notice that $b > 0$, since $o(x)$ and a are.

We have a divides m, so that $m = qa$. Then
$$(x^m)^b = (x^{qa})^b = x^{qab} = x^{q \cdot o(x)} = (x^{o(x)})^q = 1^q = 1$$

Beware that this does not complete the proof. We have merely found that when x^m is raised to the b power, the result is 1. This only implies that $o(x^m)$ *divides b*, not that they are equal – recall Corollary 7.3.

We can complete the proof by showing that b divides $o(x^m)$, for then the two natural numbers b and $o(x^m)$ will divide each other, and hence be equal. Write $j = o(x^m)$, and then $x^{mj} = (x^m)^j = 1$. By Corollary 7.3, the number $o(x)$ divides $m \cdot j$. Now we use GCD Theorem. Write $a = c \cdot m + d \cdot o(x)$, for some integers c, d, and then
$$a \cdot j = c \cdot m \cdot j + d \cdot j \cdot o(x)$$

Since $o(x)$ divides $m \cdot j$, it divides $c \cdot m \cdot j$, and obviously $o(x)$ divides $d \cdot j \cdot o(x)$. Thus $o(x)$ divides $a \cdot j$. We had $o(x) = a \cdot b$, and we have $a \cdot b$ divides $a \cdot j$. We can cancel the a's (why?), and so b divides $j = o(x^m)$. This completes the proof. $\square$

An immediate corollary.

COROLLARY 8.4. *Let x be an element of finite order in a group G and let m be an integer. Then $\langle x^m \rangle = \langle x \rangle$ if and only if m and $o(x)$ have GCD 1.*

It is a good idea to take Corollary 8.4 to some fairly simple example, computing by hand to make sure you understand what it says. This is a good review of the cyclic group $\langle x \rangle$. Although you can probably come up with your own sample x, here is a suggestion:

$$x = [123][45][6789]$$

Compute the cyclic groups generated by each power of x, and observe which of these groups is equal to which other such group. The answer should be in accord with Corollary 8.4!

Here is another application of the GCD Theorem. The linear equation $m \cdot X = b$ has solution $X = b/m$ provided that $m \neq 0$, and this works in the rational numbers, the real numbers, and the complex numbers. Let's think about the analogous equation in $\mathbb{Z}_n$. Choose a natural number n and choose integers m, b. We ask, "When does $m \cdot X \equiv b$ mod n have an integer solution X?" For instance, when $b = 1$, we are asking when m has an inverse mod n.

We claim the following answer to the question: $m \cdot X \equiv b$ mod n has a solution $X \in \mathbb{Z}$ if and only if the GCD of m, n divides b. We will prove the *only if* direction, leaving the other direction to you. Suppose $m \cdot X \equiv b$ mod n for $X \in \mathbb{Z}$. Then there is an integer c such that $m \cdot X = b + c \cdot n$. Let g be the GCD of m, n, and then g divides $m \cdot X$, and g divides $c \cdot n$. Thus, g divides b, as claimed.

8. FACTORIZATION

We see that the modular linear equation is a little more tricky than its counterpart in the real numbers. What about quadratic equations? Given $n \in \mathbb{N}$ and $a, b, c \in \mathbb{Z}$, when does $a \cdot X^2 + b \cdot X + c \equiv 0 \mod n$ have a solution $X \in \mathbb{Z}$? Quadratic equations have solutions depending on square roots. Is there an analogy in modular quadratic equations? These questions lead us into deep waters, and we are not in a position to give a solution to this problem. Gauss gave a solution to this problem in the *Disquisitiones Arithmeticae* (referred to in Chapter 3). You might be asked to experiment with special cases.

Here are two helpful facts; we will probably prove one of them in class and leave the other one to you. Both of these facts involve GCD Theorem(b).

PROPOSITION 8.5. *Let the integers a and b have GCD 1, and let c be an integer.*
(a) If a divides c and b divides c, then $a \cdot b$ divides c.
(b) If a divides $b \cdot c$, then a divides c.

The GCD has an obvious connection to the properties of primes. A *prime* is a natural number p which has exactly 2 natural numbers divisors. Thus, p is not 0 and not 1 and if $p = ab$, with $a, b \in \mathbb{N}$, then either $a = p$ or $b = p$. Notice that we have defined primes to be *positive*; thus, 3 is a prime for us but not -3. This is not essential, and the definition of *prime* in other books may include negative numbers. We will stick to positive numbers because it leads to greater simplicity in our number theoretic statements.

Primes are the building blocks of the integers.

LEMMA 8.6. *Let $n \in \mathbb{Z}$ with $n \geq 2$, Then there are primes $p_1, p_2, \ldots, p_k$, with $k \geq 1$, such that*
$$n = p_1 \cdot p_2 \cdots p_k$$
In particular, there is a prime that divides n.

PROOF. If the Lemma is false, there is a minimum integer $n \geq 2$ which cannot be factored into primes. If n is prime, then $n = p_1$ gives the required factorization; that is a contradiction.

Thus, n is not prime, and so $n = a \cdot b$ for some positive integers a, b with $1 < a < n$. It follows that $1 < b < n$. In particular, $a \geq 2$ and $b \geq 2$. By the minimality of n, the numbers a and b factor into primes, and thus n does. Contradiction. □

We note that the primes in Lemma 8.6 do not have to be distinct. For instance, $12 = 2 \cdot 3 \cdot 2$.

You probably know that there are infinitely many primes. There is a proof of this fact in Euclid's *Elements*. The proof has an unsettling effect on some people; we will discuss this in class.

THEOREM 8.7. *There are infinitely many primes.*

PROOF. If there are only finitely many primes, say there are r of them, then number them p_i for $1 \leq i \leq r$. Define

$$n = 1 + p_1 \cdot p_2 \cdot \ldots \cdot p_r$$

Clearly n is a natural number greater than 1, and so by Lemma 8.6 it has at least one prime divisor p. But p must be among the p_i, so that p divides the product $p_1 \ldots p_r$. Thus p divides the difference of n and the product of primes, which is 1. Since p divides 1, we must have $p = 1$, an impossibility. □

Now we are ready for a more subtle property of primes.

PROPOSITION 8.8. *Let a and b be integers, let p be a prime, and suppose that p divides $a \cdot b$. Then p divides a or p divides b.*

8. FACTORIZATION

PROOF. Let m be the GCD of p and a. (We know that m exists by the GCD Theorem, since p is not zero.) Since m is a natural number divisor of the prime p, we either have $m = 1$ or $m = p$.

If $m = 1$, then the GCD Theorem says that $1 = xa + yp$ for some integers x, y. Then $b = xab + ypb$. We have p dividing $a \cdot b$, and certainly p divides ypb, and therefore p divides b. (You might notice that this case could have used Proposition 8.5b.)

If $m = p$, then the definition of GCD shows that p divides a. $\square$

We are in position to sharpen Lemma 8.6. We used the example $12 = 2 \cdot 3 \cdot 2$; it is customary to collect the common primes: $12 = 2^2 \cdot 3$. The Fundamental Theorem of Arithmetic says this can be done – and in only one way. Its impressive name notwithstanding, the Fundamental Theorem is not hard to prove. The essential argument goes back at least to Euclid. And don't let all the notation put you off; this theorem just writes each integer as a sign multiplied by a product of distinct primes to powers.

FUNDAMENTAL THEOREM OF ARITHMETIC. *Let $n \in \mathbb{Z}$ with $n \neq 0$. Let P be the set of primes that divide n. Then there is a unique integer $s \in \{-1, 1\}$ and a unique function $e : P \to \mathbb{N}$ such that*

$$(8.1) \qquad n = s \cdot \prod_{p \in P} p^{e(p)}$$

(If $P = \phi$, then the product is taken to be 1.)

PROOF. If $n = 1$, then $n = 1 \cdot 1$ is the only possible product at in (8.1). If $n = -1$, then $n = -1 \cdot 1$ is the only possible product. For the rest of the proof, we assume that $|n| \geq 2$.

We have defined primes to be positive, and so the product in (8.1) is positive. Thus, s must be the sign of n, and so s is unique. Furthermore, $s \cdot n \geq 2$, and so we may as well assume that $n \geq 2$.

Lemma 8.6 shows that $n = p_1 \cdots p_k$ for various primes p_j. We claim that the set of the p_j is exactly the set P of primes that divide n. Indeed, the p_j obviously divide n, and so each is in P.

Let $q \in P$. We show that q is one of the p_j by induction on k (the number of p_j). Write $n = p_1 \cdot r$, where $r = 1$ if $k = 1$, and, when $k \geq 2$, we let r be the product of the p_j with $j \geq 2$. Then q divides $p_1 \cdot r$, and Proposition 8.8 shows that q divides p_1 or q divides r. If q divides p_1, then since p_1 and q are primes, we have $q = p_1$. If q divides r, then since r is the product of $k - 1$ primes, induction shows that $q = p_j$ for some $j \geq 2$.

Now we see how to define the function $e : P \to \mathbb{N}$. For $p \in P$ (of course $p = p_j$ for some j), define $e(p)$ to be the number of j such that $p_j = p$. Then (8.1) holds. (We have $s = 1$ since $n > 0$.)

We need to show that e is unique. Let $f : P \to \mathbb{N}$ with

$$\prod_{p \in P} p^{e(p)} = n = \prod_{p \in P} p^{f(p)}$$

If $e \neq f$, then there is $q \in P$ such that $e(q) \neq f(q)$. Suppose, without loss of generality, that $e(q) < f(q)$. Then we can cancel $q^{e(q)}$ from both sides.

$$\prod_{p \in P \setminus \{q\}} p^{e(p)} = q^{f(q)-e(q)} \cdot \prod_{p \in P \setminus \{q\}} p^{f(p)}$$

The prime q divides the right side, but not the left side. This contradicts a previous result in this proof: if a number can be written as a product of primes, then those primes are *all* the primes that divide that number. $\square$

There are many facts that follow from prime factorization. For instance, we prove that an integer prime cannot have a rational number square root. This result goes back to the Pythagoreans (before Euclid), and it has always been regarded as interesting. Without introducing rational numbers (fractions) or real numbers, we can state the result just using the integers. For if we had an equation $\sqrt{p} = m/n$, where $m, n \in \mathbb{Z}$, then we would have $n \cdot \sqrt{p} = m$, and also $n^2 \cdot p = m^2$. This last equation is an integer equation, and so it is in the domain of our course!

PROPOSITION 8.9. *Let p be a prime. Then there are no non-zero integers n and m such that $m^2 = p \cdot n^2$.*

PROOF. Assume, to the contrary that there are non-zero integers n and m such that $m^2 = p \cdot n^2$. We may as well assume that m and n are positive. Use the Fundamental Theorem of Arithmetic; let P be the primes dividing n and let Q be the primes dividing m, and then

$$m = \prod_{q \in Q} q^{f(q)} \quad \text{and} \quad n = \prod_{q \in P} q^{e(q)}$$

So that $m^2 = p \cdot n^2$ is this:

$$\prod_{q \in Q} q^{2 \cdot f(q)} = p \cdot \prod_{q \in P} q^{2 \cdot e(q)}$$

Because p divides the left side, the Fundamental Theorem says $p \in Q$. The exponent of p on the left is $2 \cdot f(p)$, which is even; the exponent on the right is $1 + 2 \cdot e(p)$, which is odd. The uniqueness of the exponents (the functions e, f) is thereby contradicted. $\square$

There are many excellent books on elementary number theory, for instance *Elementary Number Theory* by D. Burton, Allyn and Bacon 1976; and *Elementary number theory and its applications* by K. Rosen, Addison-Wesley 1984. Many of the results in this section are in Euclid's Books VII-IX of *The Elements, Vol. 2,* by Euclid, (English translation, Dover, 1956). For instance, Euclid's algorithm for the greatest common factor is Proposition 2 of Book VII, and Euclid's proof that there are infinitely many primes is Proposition 20 of Book IX. You can and should read the masters!

Problems

95. Prove Proposition 8.1: that integers a, b have GCD equal to 1 if and only if there are integers c, d such that $a \cdot c + b \cdot d = 1$.

96. Let $a, b \in \mathbb{Z}$ and $n \in \mathbb{N}$ with $a \equiv b$ mod n. Then the GCD of a, n is the GCD of b, n.

97. Let $n \in \mathbb{N}$ and $x \in \mathbb{Z}_n$. Show that x has order n (with addition as operation) if and only if $x \in U_n$.

98. Let $p \in \mathbb{N}$ be a prime. Show that $|U_p| = p - 1$ and that $|U_{p^2}| = p \cdot (p - 1)$. (Hint for p^2: write each element of $\mathbb{Z}_{p^2}$ as $q \cdot p + r$ as in the Division Theorem.)

99. Prove one direction of Proposition 8.2: if $x \in U_n$, then x, n have GCD 1.

100. Find the order of each element of U_{13}. Choose a generator of this group, and show that Corollary 8.4 holds.

101. Find an integer x with $65 < x < 200$ and such that x has order 20 as an element of $\mathbb{Z}_{200}$ (under addition).

102. Let $m \in \mathbb{Z}$ and $n \in \mathbb{N}$, and suppose that the GCD of m, n divides the integer b. Prove that the equation $m \cdot X \equiv b$ mod n has an integer solution X.

103. Prove Proposition 8.5(a). (Hint: GCD Theorem(b).)

104. Prove Proposition 8.5(b).

105. Let F be the set of natural numbers having an even number of natural number factors. Look at $n \in \mathbb{N}$ with $n \leq 50$, and decide which of these n is in F. Make a conjecture of the form "$n \in F$ if and only if ...." (This is a computational problem – you're not being asked for a proof, but do show your work.)

8. FACTORIZATION

106. Let $a, b \in \mathbb{N}$ and assume that the GCD of a, b is 1. Suppose that $a \cdot b$ is the square of an integer. Show that a is the square of an integer. (Hint: Fundamental Theorem applied to a and to b.)

107. Let $a, b \in \mathbb{Z}$ have GCD 1. Show that $a - b$ and $a + b$ have GCD 1 or 2, and give examples to show that each case can actually occur.

108. Let G be a finite group, let $x \in G$, and let m be an integer, and suppose that the GCD of m and $o(x)$ is 1. Show that there is $y \in \langle x \rangle$ such that $y^m = x$.

109. A natural number is *square-free* if it that has no divisor, other than 1, that is an integer square. Show that a natural number greater than 1 is square-free if and only if it is the product of distinct primes to the first power. (Hint: Fundamental Theorem.)

110. Let $n \geq 2$ be a square-free natural number. Show that there are no non-zero integers x, y such that $x^2 = n \cdot y^2$.

111. Let a, b, c be positive integers such that $a^2 + b^2 = c^2$ and a, b have GCD 1. (We say that a, b, c are a *Pythagorean triple*.) A problem on p.28 shows that one of a, b is even and the other odd, and c is odd. Say that a is even. Follow the steps to prove that there are integers m, n such that $c = m^2 + n^2$ and $a = 2 \cdot m \cdot n$ and $b = m^2 - n^2$.

(a) Show that a, c have GCD equal to 1. (Hint: let p be a prime dividing a, c.)
(b) Use $a^2 = (c - b) \cdot (c + b)$ to show that $c - b$ and $c + b$ have GCD 2.
(c) Show that there are integers m, n such that $c - b = 2 \cdot n^2$ and $c + b = 2 \cdot m^2$. Then also, $a^2 = 4 \cdot m^2 \cdot n^2$. (Hint: $c - b$ and $c + b$ are even. Besides the factor of 2, suppose one of them had a prime p to an odd power. Show that prime would have to occur in the other factor, as well. But it can't.)
(d) Show that $c = m^2 + n^2$ and $b = m^2 - n^2$ and $a = 2 \cdot m \cdot n$.

112. Let $n \geq 3$ be an integer. Show that $\mathbb{Z}_{2^n}$ has exactly four elements whose square is 1. (Hint: for $c \in \mathbb{Z}_{2^n}$, think about the power of 2 that divides $c-1$ and $c+1$.)

113. Find the smallest odd natural number n such that $\mathbb{Z}_n$ has more than two elements x with $x^2 \equiv 1$.

114. Find the smallest odd natural number n such that $\mathbb{Z}_n$ has at least two elements x with $x^5 \equiv 1$.

CHAPTER 9

Subgroups

For an element x of a group G, the subset $\langle x \rangle$ of G is itself a group under the same operation as G. We have also seen how D_4 is a subset of D_8, and these two groups share the same operation (function composition). The group of symmetries of a graph is a subgroup of the group of permutations on the vertices of the graph, and these groups share function composition as operation. In general, if H is a subset of a group G, and if H is itself a group under the operation of G, then we say that H is a *subgroup* of G. If H is not all of G, we call H a *proper subgroup* of G.

It is obvious and trivial that every group is a subgroup of itself and that the set consisting just of the identity element of a group is a subgroup of the group. As we mentioned above, if G is a group and $x \in G$, then $\langle x \rangle$ is a subgroup of G.

It is an exercise to show that the alternating group $\mathbb{A}_P$ is a subgroup of $\mathbb{S}_P$.

You will prove that the intersection of two subgroups is a subgroup, and that the union of two subgroups is never a subgroup (unless that union is equal to one of the two subgroups used in the union, in which case the union is not interesting).

There is one further triviality to mention: if K is a subgroup of H and H is a subgroup of G, then K is a subgroup of G: a subgroup of a subgroup is a subgroup!

Given a subset H of a group G, how do we check whether H is a subgroup? The fact that the operation of G must be an operation on H tells us that if $x, y \in H$ then $xy \in H$ too. (We say that H is *closed under the operation*.) Looking at the other group axioms, the operation on H must be associative, but since the operation is already associative on G (since G is a group!), we get the associative law on H for free. Next, there must be an identity element in H. Since Proposition 5.2 says that there can be only one identity element in all of G, we must have 1 (the identity of G) must be in H. Finally, each element of H must have an inverse. Of course they do, you say, since G is a group. But these inverses must be elements of H, since it is H that is trying to be a group!

Let us give a very efficient summary of the previous paragraph, answering the question, "How do we show that a subset H of a group G is, in fact, a subgroup?" Since we will assume that G is finite, the answer is very simple: if H is non-empty and closed under the operation, it is a subgroup.

PROPOSITION 9.1. *Let H be a subset of a finite group G. Then H is a subgroup of G if and only if H is non-empty, and if $x, y \in H$, then $x \cdot y \in H$.*

PROOF. If H is a subgroup of G, then H is nonempty, since $1 \in H$. If $x, y \in H$, then, $x \cdot y \in H$, since H is a subgroup.

Now assume that H is non-empty and closed under the group operation. This shows that the operation of G is an operation on H. Since H is nonempty, there is an element $x \in H$. Then $x, x \in H$, and so $x^2 = x \cdot x \in H$. Again, since $x, x^2 \in H$, we have $x^3 \in H$. Continuing, we see that $x^k \in H$ for all positive integers k. In particular, $1 = x^{o(x)} \in H$, and so H has the identity element, and that element is its own inverse.

If $x \neq 1$, then $o(x) \geq 2$, and so $x^{-1} = x^{o(x)-1} \in H$. Thus, the elements of H have inverses in H. We see that H is a subgroup of G. □

If G is a group, we know that $\{1\}$ and G are (uninteresting) subgroups of G. It is interesting, however, to ask what groups have no *interesting* subgroups: for which groups G are $\{1\}$ and G the *only* subgroups?

Here are some typical subgroups. The proof of the following will be an exercise. Pay close attention to the quantifiers in these definitions.

PROPOSITION 9.2. *Let G be a group, and let $x \in G$. Then*
$$C_G(x) = \{y \in G \mid xy = yx\}$$
is a subgroup of G. Also, the set
$$Z(G) = \{z \in G \mid zy = yz \text{ for all } y \in G\}$$
is an abelian subgroup of G.

For $x \in G$, the group $C_G(x)$ is the *centralizer of x in G*. You will find the centralizer of an n-cycle in $\mathbb{S}_n$ and the centralizer of a 2-cycle in that group.

The group $Z(G)$ is called the *center* of the group G. What is the center of D_8? Of D_6? Of Q_8? Of $\mathbb{Z}_n$?

Here is another example: if P is a finite set and $a \in P$, then the set of all $f \in \mathbb{S}_P$ such that $f(a) = a$ is a subgroup of $\mathbb{S}_P$. (These are the permutations of P that have a as a fixed point.)

Recall that a group G is cyclic if it is generated by some one of its elements: $G = \langle x \rangle$ for some $x \in G$. Observe that $\mathbb{Z}_n$ is cyclic, since $\mathbb{Z}_n = \langle 1 \rangle$. We can describe all the subgroups of a cyclic group $G = \langle x \rangle$. All the subgroups are cyclic, and there is exactly one of order m for each m that divides $o(x)$.

PROPOSITION 9.3. *Let $G = \langle x \rangle$, with $o(x) = n$.*

(a) Let H be a subgroup of G. Then H is cyclic and $|H|$ divides n.

(b) Let q be a natural number dividing n. Then there is a unique subgroup of G of order q.

PROOF. Let H be a subgroup of $\langle x \rangle$. We need to show that H is cyclic, in other words that $H = \langle y \rangle$ for some $y \in H$. Since $y \in \langle x \rangle$, we would have $y = x^m$ for some m. Observe that the elements of $\langle y \rangle$ would be powers of x^m; this suggests how we might find such a y. Define
$$S = \{m \in \mathbb{N} \mid x^m \in H\}$$
Since $1 \in H$, and $1 = x^n$, we see that S is not empty. Let m be the minimal element of S, and we claim that $H = \langle x^m \rangle$. Indeed, the definition of S shows that $x^m \in H$, so that $\langle x^m \rangle$ is a subgroup of H.

Let $y \in H$. Then $y \in \langle x \rangle$, and so $y = x^k$ for some integer k. The Division Theorem produces $k = mq + r$ where $0 \le r < m$. Then
$$x^k = x^{mq+r} = (x^m)^q \cdot x^r$$
so that
$$x^r = (x^m)^{-q} \cdot x^k$$
is seen to be the product of two elements of H, hence $x^r \in H$. If $r > 0$, then this puts r in S, but then $r < m$ contradicts the minimality of m. Thus $r = 0$, whence $k = mq$ and
$$x^k = (x^m)^q \in \langle x^m \rangle$$
We conclude that H is a subgroup of $\langle x^m \rangle$. The two sets H and $\langle x^m \rangle$ each contain the other, and so they are equal. In particular, H is cyclic.

Next we claim that m divides n. Indeed, the Division Theorem writes $n = qm + r$ where $0 \le r < m$. Then
$$1 = x^n = x^{qm} \cdot x^r \quad \text{so that} \quad \left(x^{mq}\right)^{-1} = x^r$$
This proves that $x^r \in H$. If $r > 0$, then $r < m$ contradicts the minimality of m; thus, $r = 0$ and m divides n.

If we write $n = ma$, then it is easy to see that there are a powers of x^m, and so
$$H = \{x^m, x^{2m}, \cdots, x^{ma} = 1\}$$

Thus, $|H| = a$ divides n. Statement (a) is proved.

For (b), let q be a natural number dividing n, and write $n = qm$. As we just saw, the order of x^m is q, and so $\langle x^m \rangle$ is a subgroup of G of order q. To complete the proof, we need to show that this subgroup is unique.

Let H be a subgroup of G with $|H| = q$. By the proof for (a) we know that $H = \langle x^k \rangle$ where k divides n, and when $n = ka$, we have that $|H| = a$. Since $|H| = q$, we see that $a = q$, and so $k = m$, whence $H = \langle x^m \rangle$, as needed. □

There is an interesting number-theoretic consequence of the preceding; to get to it, we need some discussion. For a natural number n as modulus, and for $x \in \mathbb{Z}_n$, we showed previously that all $y \equiv x$ have the same GCD with n. Thus, we can speak of the GCD of x and n, thinking of x as an element of $\mathbb{Z}_n$.

Let n be a positive integer, and recall the set U_n of $x \in \mathbb{Z}_n$ such that x and n have GCD 1. We are interested in the order of U_n. This number is important in its own right; it is called the *totient function* and denoted $\phi(n)$ (another of Euler's ideas). You might compute some values: $\phi(1) = 1$, $\phi(2) = 1$, $\phi(3) = 2$, $\phi(24) = 8$. If you recall the original definition of U_n, you see that $\phi(n)$ is the number of elements $m \in \mathbb{Z}_n$ that have a multiplicative inverse. A homework problem showed that if p is prime, then $\phi(p) = p - 1$ and $\phi(p^2) = p \cdot (p - 1)$.

Let $\langle x \rangle$ be a cyclic group of order n. By Proposition 7.2, the elements of $\langle x \rangle$ look like x^r where $0 \leq r < n$, and by Corollary 8.4, the order of x^r is n if and only if the GCD of r and n is 1. In other words, the number of elements of $\langle x \rangle$ having order n is $\phi(n)$.

Now we are ready for a number-theoretic consequence of group theory.

THEOREM 9.4. *Let n be a natural number, and define V to be the set of natural number divisors of n. Then*

$$\sum_{m \in V} \phi(m) = n$$

PROOF. Let $\langle x \rangle$ be a cyclic group of order n (such a group exists, for example $\mathbb{Z}_n$). Proposition 8.3 shows that the orders of elements of $\langle x \rangle$ are natural number divisors of $o(x) = n$. Thus, for each natural number m which divides n (i.e. each $m \in V$), define $f(m)$ to be the number of elements of $\langle x \rangle$ of order m. Then,

$$(9.1) \qquad \sum_{m \in V} f(m) = n$$

since this formula merely categorizes elements of $\langle x \rangle$ by their orders.

We will complete the proof by showing that $f(m) = \phi(m)$. Look at some particular natural number m dividing n. Proposition 9.3b provides an element y of $\langle x \rangle$ of order m. If z is an element of $\langle x \rangle$ of order m, then $\langle z \rangle$ is a subgroup of $\langle x \rangle$ of order m. Proposition 9.3b tells us that there is only one such subgroup, therefore $\langle z \rangle = \langle y \rangle$, so that $z \in \langle y \rangle$. We see that the elements of $\langle x \rangle$ of order m are the elements of $\langle y \rangle$ of order m. How many such elements are there in $\langle y \rangle$? The paragraph before this theorem shows that there are exactly $\phi(m)$ elements of $\langle y \rangle$ order m. Hence $\langle x \rangle$ has exactly $\phi(m)$ elements of order m. We have proved that $f(m) = \phi(m)$, and the conclusion of this theorem now follows from equation (9.1). □

Problems

115. If H, K are subgroups of the group G, then $H \cap K$ is a subgroup of G.

116. Suppose that H and K and $H \cup K$ are subgroups of a group G. Show that $H \subseteq K$ or $K \subseteq H$. (Hint: assume that $H \not\subseteq K$ and $K \not\subseteq H$.)

117. Let H and K be subgroups of the abelian group G. Define HK to be the set of all elements $h \cdot k$, where $h \in H$ and $k \in K$. Show that HK is a subgroup of G.

118. Show that $\langle[12]\rangle\langle[13]\rangle$ is not a subgroup of $\mathbb{S}_3$. (Note: use the same definition as HK in the previous problem.)

119. Let G be a group and $x \in G$. Show that $C_G(x)$ is a subgroup of G. (Definition in Proposition 9.2.)

120. Prove that the center $Z(G)$ of a group G is a subgroup of G. (Definition in Proposition 9.2.)

121. Show that the center of $\mathbb{S}_n$ consists only of the identity element when $n \geq 3$.

122. Let p be an integer prime, and let k be a positive integer. Show that $\phi(p^k) = p^{k-1} \cdot (p-1)$.

123. Let P be a set, choose $a \in P$ and define H to be the set of $h \in \mathbb{S}_P$ such that $h(a) = a$. Show that H is a subgroup of $\mathbb{S}_P$.

124. Let x, y be elements of a finite group G, and suppose that $o(x)$ and $o(y)$ have GCD equal to 1. Suppose also that $xy = yx$. Show that the order of xy is $o(x) \cdot o(y)$.

125. For groups G, H the definition of the group $G \times H$ is on p.64.
(a) Show that $Z(G \times H) = Z(G) \times Z(H)$
(b) For $g \in G$ and $h \in H$ show that $C_{G \times H}(g, h) = C_G(g) \times C_H(h)$.

126. Find the subgroups of $\mathbb{S}_3$. (Hint: there are six of them.)

127. Find the subgroups of D_8.

CHAPTER 10

Lagrange's Theorem

In all the examples we have considered of a subgroup H in a group G, we have $|H|$ divides $|G|$. We wish to prove that the order of a subgroup always divides the order of the (finite) group containing it. Our goal, Lagrange's Theorem, is regarded as the starting point for the serious study of groups. It was proved in the late 1700's.

The proof of Lagrange's Theorem entails an important new idea, that of the *coset*. To motivate this idea, we consider the kind of specific example that originally interested Lagrange.

Let A be a finite set, and consider the group $\mathbb{S}_A$. Suppose that $a, b \in A$, and define

$$F(a \to b) = \{f \in \mathbb{S}_A \mid f(a) = b\}$$

Notice that $F(a \to a)$ was considered at the end of Chapter 9; it is a subgroup of $\mathbb{S}_A$. The other sets $F(a \to b)$ are not subgroups, but Lagrange saw how the various $F(a \to b)$ are related to the subgroup $F(a \to a)$. The relationship between $F(a \to b)$ and $F(a \to a)$ is seen by choosing some particular element f of $F(a \to b)$ (an obvious example is $f = [ab]$). We claim that $h \in F(a \to b)$ if and only if $h = fg$ for some $g \in F(a \to a)$. We have fixed f, and so the last sentence relates $h \in F(a \to b)$ to $g \in F(a \to a)$.

To prove the claim, let $g \in F(a \to a)$, and then $fg(a) = f(a) = b$, so that $fg \in F(a \to b)$. Conversely, let $h \in F(a \to b)$, and define $g = f^{-1}h$. (In other words, make sure $h = fg$.) We need to show that $g \in F(a \to a)$.

We already have $h(a) = b$, and since $f(a) = b$, we see that $f^{-1}(b) = a$; thus, $f^{-1}h(a) = f^{-1}(b) = a$, so that $g(a) = a$ as needed.

It might help to do a very specific example. In $\mathbb{S}_4$, consider the relationship between $F(4 \to 4)$ and $F(4 \to 3)$. We have

$$F(4 \to 4) = \{E, [12], [13], [23], [123], [132]\}$$
$$F(4 \to 3) = \{[143], [1243], [34], [1432], [12][34], [243]\}$$

Choose an element f from $F(4 \to 3)$ (how about [143]), and you should compute directly that every element h of $F(4 \to 3)$ looks like $[143]g$ where $g \in F(4 \to 4)$.

Notice that the correspondence from $F(a \to a)$ to $F(a \to b)$ suggests that these sets have the same size. We call $F(a \to b)$ a *coset* of the subgroup $F(a \to a)$. Here is the general definition for arbitrary groups and subgroups. If G is a group, and H is a subgroup of G, and $x \in G$, then

$$xH = \{xh \mid h \in H\}$$

The set xH is a *left coset* of H in G. (It is a *left* coset because the x is on the left of the subgroup; there are also *right cosets* Hx defined analogously, we will not need them right now.) Here are the properties of left cosets that mimic those of $F(a \to b)$ above.

PROPOSITION 10.1. *Let H be a subgroup of the group G, and let $x \in G$. Then the function from H to xH defined by sending $h \in H$ to xh, is one to one and onto. If G is finite, then $|xH| = |H|$. If $x, y \in G$, then the sets xH and yH are either equal or disjoint.*

PROOF. It is easy to show that the function defined in the statement of the proposition is one to one and onto, and this proves the first two conclusions.

Let xH and yH have at least one element in common. This element can be labelled xh for some $h \in H$, or yk for $k \in H$. Thus $xh = yk$, or $x = ykh^{-1}$. Let us show that $xH \subseteq yH$. An element of xH can be written xg for some $g \in H$. Then $xg = ykh^{-1}g$, and since k, h, g are all in the subgroup H, we see that $kh^{-1}g \in H$, whence $xg \in yH$. We have shown that $xH \subseteq yH$. Interchanging x and y shows that $yH \subseteq xH$. □

The set of cosets of H in G is denoted G/H. If G is finite, then G/H is a finite set.

LAGRANGE'S THEOREM. *Let G be a finite group and H a subgroup of G. Then G is the disjoint union of the distinct left cosets of H in G. Also $|G| = |H| \cdot |G/H|$ so that $|H|$ divides $|G|$.*

PROOF. Each $x \in G$ is in the coset xH, for we have $x = x1$ and $1 \in H$. Thus every element of G is in a coset. Proposition 10.1 says that distinct cosets are disjoint, and so each element is in exactly one coset. This proves that G is the disjoint union of the cosets.

By Proposition 10.1, each coset has size $|H|$. There are $|G/H|$ cosets, they are disjoint, and they cover G. Thus, $|G| = |H| \cdot |G/H|$ as needed. □

Lagrange's Theorem has many consequences. Here is one such.

COROLLARY 10.2. *Let x be an element of a finite group G. Then $o(x)$ divides $|G|$.*

PROOF. The number $o(x)$ is the order of the subgroup $\langle x \rangle$, by Proposition 7.2. □

Here is a consequence of Corollary 10.2. You will need to recall Theorem 9.4.

PROPOSITION 10.3. *Let G be a finite group, and suppose, for each positive integer k such that G has an element of order k, we have that G has exactly k elements x with $x^k = 1$. Then G is cyclic.*

PROOF. Let $n = |G|$. We intend to partition the elements of G by their various orders. If $x \in G$, then by Corollary 10.2 the order $o(x)$ is a divisor of n; we also know that $o(x)$ is a positive integer. Thus, if we let V be the set of positive integer divisors of n, and if, for each $k \in V$, we let $V(k)$ be the set of elements of G having order k, then

$$(10.1) \qquad n = |G| = \sum_{k \in V} |V(k)|$$

(Of course, some of the $V(k)$ may be empty.)

On the other hand, Theorem 9.4 yields

$$(10.2) \qquad n = \sum_{k \in V} \phi(k)$$

Subtracting equation (10.1) from equation (10.2) we obtain

$$(10.3) \qquad 0 = \sum_{k \in V} (\phi(k) - |V(k)|)$$

We intend to show, for each $k \in V$, that $\phi(k) \geq |V(k)|$. Before we do this, let us see how it will complete the proof. The inequality claimed would show that the terms $\phi(k) - |V(k)|$ of equation (10.3) are non-negative numbers. The only way a sum of non-negatives can be 0 is if each of the numbers is 0. In other words, we would know that $\phi(k) = |V(k)|$ for all divisors k of n. Taking the case $k = n$ we see that $\phi(n) = |V(n)|$; what we want from this is merely that $|V(n)| \neq 0$, so that $V(n)$ is not empty. Going back to the definition of $V(n)$, this shows that there is an element x of the group G of order n. But then $\langle x \rangle$ would be a subgroup of G of order n, and since $|G| = n$, we see that $G = \langle x \rangle$, so that G is cyclic, and the Theorem is proved.

We need to show, for $k \in V$, that $\phi(k) \geq |V(k)|$. The inequality is certainly true if $|V(k)| = 0$, and so we assume that $|V(k)| > 0$. This produces an element y of G having order k. The subgroup $\langle y \rangle$ then has order k, and all of its elements $z \in \langle y \rangle$ satisfy $z^k = 1$. The hypothesis on G says that no more than k elements can have k-th power equal to 1, and we conclude that the elements of $\langle y \rangle$ are precisely the elements of G whose k-th power is 1.

Now we can show that $|V(k)| = \phi(k)$. Given $z \in V(k)$, we have $z^k = 1$. The previous paragraph then proves that $z \in \langle y \rangle$. The discussion before Theorem 9.4 showed that the cyclic group $\langle y \rangle$ of order k has exactly $\phi(k)$ elements of order k. The element z must be one of these elements. This proves that $|V(k)| = \phi(k)$, and again the desired inequality holds. □

Lagrange's Theorem says that the order of a subgroup divides the order of the group. Suppose we have a natural number that divides the order of a group; is there a subgroup of that order? Maybe not. The smallest example occurs in the alternating group on four points $\mathbb{A}_4$; that group has order 12 and it has no subgroup of order 6. This fact follows from a direct calculation that can be shortened by more advanced results. It might be worth doing now, however. A hint is provided in the problems.

Cosets can be tricky. Suppose that two cosets xH and yH of H is a subgroup of G are equal:
$$xH = yH$$
It is tempting to "cancel the H's," concluding that $x = y$. This is false, and the truth is complicated but under control. We will prove the following in class.

PROPOSITION 10.4. *Let H be a subgroup of the group G. Let $x, y \in G$. Then the following are equivalent:*

(a) $xH = yH$;
(b) $x \in yH$;
(c) $y^{-1}x \in H$;

Our next use of cosets is to count the elements of the set HK where H, K are subgroups of a group G. The definition is natural and has occurred in two

previous problems.
$$HK = \{hk \mid h \in H, k \in K\}$$
Sometimes this set is a subgroup of G and sometimes it isn't. (The trouble is that it is not always closed under the group operation.) Notice that $HH = H$ for a subgroup H, and, if K is a subgroup of H, then $HK = KH = K$. In general, HK and KH are not equal.

We can always find the order of HK, regardless.

PROPOSITION 10.5. *If H and K are subgroups of the finite group G, then*
$$|HK| \cdot |H \cap K| = |H| \cdot |K|$$

PROOF. We first show that HK is the union of left cosets hK for $h \in H$. Indeed, if $h \in H$ and $k \in K$, then $hk \in hK$, and that does it.

Proposition 10.1 shows that distinct cosets are disjoint; if HK involves d cosets hK, then $|HK| = d \cdot |K|$.

Let $L = H \cap K$, a subgroup of G. We will show that $d = |H/L|$. This will complete the proof, for Lagrange's Theorem shows that $d \cdot |L| = |H|$, and if we multiply the equation $|HK| = d \cdot |K|$ by $|L|$, the conclusion we want will follow.

To show that d is the number of cosets of L in H, we look at a left coset hL for $h \in H$. We claim that $xK = hK$ for every $x \in hL$. Indeed, such an x can be written hy for $y \in L$, and then $xK = hyK$. Since $y \in L \subseteq K$, we have that $yK = K$ by Proposition 10.4. Thus, $xK = hK$, as claimed. In other words, every $x \in hL$ gives the same coset $xK = hK$ of K. We can therefore define a function from H/L to the cosets of K in HK, sending the coset hL to the coset hK.

This function is onto, since the coset hK comes from hL. The function is also one to one, since if $hK = yK$ for $h, y \in H$, then Proposition 10.4 shows that $y^{-1}h \in K$ and we already have $y^{-1}h \in H$, so that $y^{-1}h \in L$.

Proposition 10.4 then says that $hL = yL$. Since our function is one to one and onto, the number d of left cosets of K in HK is equal to the number of cosets of L in H. □

Lagrange's Theorem shows that the order of $H \cap K$ divides both $|H|$ and $|K|$. If the latter two integers have GCD equal to 1, then $|H \cap K| = 1$, and so Proposition 10.5 shows that $|HK| = |H| \cdot |K|$ in this case.

Problems

128. Find the cosets of $\langle [12] \rangle$ in $\mathbb{S}_3$. Find the cosets of $\langle 8 \rangle$ in $\mathbb{Z}_{12}$.

129. Show that $\mathbb{A}_4$ has no subgroup of order 6. (Hint: Since 6 is even, the subgroup has an element of order 2. What else can be there?)

130. Let H be a subgroup of the finite group G, and let K be a subgroup of H. Show that $|G/K| = |G/H| \cdot |H/K|$.

131. Suppose that H is a subgroup of the group G and that $|G/H|$ is prime. Suppose that K be a subgroup of G with $H \subseteq K$. Show that $K = H$ or $K = G$.

132. Let G be a finite group. Show that G has *exactly* two subgroups if and only if $|G|$ is prime. (Hint: ⇒: show that G is cyclic and use Proposition 9.3; ⇐: Lagrange.)

133. Let p be a prime and suppose that G is a group of order p^e, for some $e \geq 1$. Show that G has an element of order p.

134. Let G be a group of order 55. Show that G has an element of order 11 and an element of order 5. (Hint: if there is no element of order 11, what is the order of all non-identity elements of G? Show that $G \setminus \{1_G\}$ divides up into disjoint subsets $\langle x \rangle \setminus \{1_G\}$, where x has order 5.)

135. Let G be a group of order $p \cdot q$ where $p < q$ are distinct primes and $p - 1$ does not divide $q - 1$. Show that G has elements of order p and order q. (Hint: the argument of the previous problem.[1])

136. Let A_1, A_2 be groups such that the GCD of $|A_1|$ and $|A_2|$ is 1. Let H be a subgroup of the direct product $A_1 \times A_2$. Define H_1 to be the set of $a \in A_1$ such that there is $b \in A_2$ with $(a, b) \in H$. Show that H_1 is a subgroup of A_1. Similarly, define H_2 as a subgroup of A_2. Show that $H = H_1 \times H_2$. (Hint: obviously, $H \subseteq H_1 \times H_2$. Let $a \in H_1$. Get $c \in H_2$ such that $(a, c) \in H$. Show that $o(a)$ and $o(c)$ have GCD 1, and so, $\langle a^{o(c)} \rangle = \langle a \rangle$. Show $(a, 1_{A_2}) \in H$. Similarly, $(1_{A_1}, b) \in H$, for all $b \in H_2$. Conclude that $H = H_1 \times H_2$.)

---

[1] We will see later that a group of order $p \cdot q$, for distinct primes p, q has elements of order p, q regardless of whether $p - 1$ divides $q - 1$.

CHAPTER 11

Normal Subgroups

We have considered the expression HK, where H, K are subgroups of a group. We can consider the similar expression $S \circ T$, where S and T are arbitrary *subsets* of a group G:

(11.1) $$S \circ T = \{s \cdot t \mid s \in S, t \in T\}$$

This defines $\circ$ as an operation on the subsets of G, and you should verify that $\circ$ is associative. This does not make the set of subsets of G into a group, for although there is an identity element (what is it?), not all subsets have an inverse (which ones don't?). When $x \in G$ and H is a subgroup of G, we have $\{x\} \circ H = xH$ (the coset!), and when H, K are subgroups of G, we have $H \circ K = HK$. For instance, $H \circ H = H$.

Recall the symmetry group D_8 with its rotation $[1234]$. Let $H = \langle [1234] \rangle$, so that $|H| = 4$ and $|D_8/H| = 2$. The cosets of H in D_8 are: H, itself (the set of rotations) and the set K of the other elements of D_8 (the set of reflections). We can compute that:

$$H \circ H = H \qquad H \circ K = K$$
$$K \circ H = K \qquad K \circ K = H$$

We see that the set $D_8/H = \{H, K\}$ forms a group under the operation of set multiplication. The young Galois discovered general examples with this same property around 1830, and his pursuit of it constituted a fundamental advance in the study of groups, an advance which became a guiding principle in much of algebra. Indeed, the investigation on which we are about to embark led Galois to the solution of a centuries' old problem regarding roots of polynomials. Although we will not now be able to discuss this problem, we wish to make clear the importance of the topic.

Let H be a subgroup of a group G. It is not hard to see that the product $xH \circ yH$ of elements of G/H is a *union* of cosets of H. If this product is always a *single* coset, then $\circ$ is an operation on G/H, as in the case above.

If you let $L = \langle [14][23] \rangle \subset D_8$, let $R = [1234]$, and observe that $RL \circ RL$ is **not** a single coset of L, and so $\circ$ is not an operation on D_8/L. In other words, the product $\circ$ is not always an operation on G/H, where H is a subgroup of the group G.

Undaunted, we pursue the question in general. In order for $\circ$ to be an operation, we need that for all $x, y \in G$ there is some $z \in G$ such that

(11.2) $$xH \circ yH = zH$$

The element $x1y1 = xy$ is an element of the product on the left side of (11.2), and so we will need xy to be an element of zH. But then Proposition 10.4 tells us necessarily that

$$zH = xyH$$

Thus, the only way for (11.2) to hold is to have

$$xH \circ yH = xyH \quad \text{for all} \quad x, y \in G$$

Let us now state precisely when this happens.

PROPOSITION 11.1. *Let H be a subgroup of a group G. Then the following are equivalent.*

(a) Every product of cosets of H is a coset of H.
(b) $xH \circ yH = xyH$ for all $x, y \in G$.
(c) $xhx^{-1} \in H$ for all $x \in G$ and $h \in H$.
(d) $xH = Hx$ for all $x \in G$.

PROOF. We have already proved that (a) implies (b).

Assume that (b) holds. For $x \in G$ and $h \in H$, we have xhx^{-1} is an element of $xH \circ x^{-1}H$. By (b), this is the coset $xx^{-1}H = H$. Thus, $xhx^{-1} \in H$, and (c) holds.

Assume that (c) holds. An element of Hx looks like hx for some $h \in H$. Write $hx = x \cdot x^{-1}hx$. Applying (c) to x^{-1} in place of x, we see that $x^{-1}hx \in H$. Thus, $hx \in xH$, and we have proved that $Hx \subseteq xH$. For the converse, an element of xH is xh for some $h \in H$. Write

$$xh = xhx^{-1} \cdot x$$

and (c) shows that $xh \in Hx$. The sets xH and Hx are equal.

Finally, assume (d). If $x, y \in G$, then

$$xH \circ yH = xHyH = x(Hy)H = x(yH)H = xyHH = xyH$$

$\square$

If H is a subgroup of G for which any of Proposition 11.1(a,b,c,d) hold, then we call H a *normal subgroup* of G, and we write $H \triangleleft G$. The word "normal" is not a great choice, since normal subgroups are rare, in the main, but this word survives as a tribute to Galois' (French) terminology.

Let P be a finite set with at least 2 elements; we will prove that $\mathbb{A}_P$ is a normal subgroup of $\mathbb{S}_P$. This proof is suggestive in a way that will be explained in the next chapter. Recall the parity function σ on $\mathbb{S}_P$, and remember that $y \in \mathbb{A}_P$ if and only if $\sigma(y) \equiv 0$. Let $x \in \mathbb{S}_P$, and use statement (d) of the Parity Theorem to compute in $\mathbb{Z}_2$.

$$\sigma(x \cdot y \cdot x^{-1}) \equiv \sigma(x) + \sigma(y) + \sigma(x^{-1}) \equiv \sigma(x) + \sigma(x^{-1})$$

By an exercise in Chapter 5, we have $\sigma(x) \equiv \sigma(x^{-1})$, and we see that

$$\sigma(x \cdot y \cdot x^{-1}) \equiv 2 \cdot \sigma(x) \equiv 0$$

This proves that $x \cdot y \cdot x^{-1} \in \mathbb{A}_P$. Proposition 11.1c then shows that $\mathbb{A}_P \triangleleft \mathbb{S}_P$.

Proposition 9.2 defined the center of a group; this is always a normal subgroup of the group.

You should see that every subgroup of an abelian group is normal.

If $G = N \times K$, a direct product of groups N and K, then the set of all $(n, 1)$ such that $n \in N$ is a normal subgroup of G. Similarly, the set of all $(1, k)$ is a normal subgroup of G.

The following is a re-statement of Proposition 11.1 and the discussion that preceded it. It summarizes what we want out of normal subgroups in general.

THEOREM 11.2. *Let H be a normal subgroup of the group G. Then G/H forms a group under the operation $\circ$, where $xH \circ yH = xyH$ for all $x, y \in G$. The identity element of G/H is H, and $(xH)^{-1} = x^{-1}H$.*

The group indicated in Theorem 11.2 is called the *quotient group* of G over H. You must keep in mind that the *elements* of the quotient group are *sets of elements* of G (cosets!). Unfortunately, the word "order" now becomes doubly ambiguous. When we speak of the "order of xH" do we mean the number of elements: $|xH| = |H|$, or do we refer to the order $o(xH)$ of the element xH of G/H? The two numbers are rarely the same. Small comfort may be derived from the fact that the particular meaning of "order" ought always to be clear from context. Watch out!

When we introduced $\mathbb{Z}_n$ in Chapter 3, we deferred the question of a more formal definition. The quotient group provides such a definition. The integers form an abelian group under addition, and if $n \in \mathbb{N}$, then $n \cdot \mathbb{Z}$ (the set of multiples of n) is a normal subgroup. Therefore, the quotient group $\mathbb{Z}/(n\mathbb{Z})$ is defined. A coset of $n\mathbb{Z}$ would look like $a + n\mathbb{Z}$ for $a \in \mathbb{Z}$. This coset is the set of $a + n \cdot b$ for all $b \in \mathbb{Z}$. In other words, the coset is the set of $m \in \mathbb{Z}$ such that $m \equiv a \mod n$. Being in the coset $a + n\mathbb{Z}$ is the same as being equal to a in $\mathbb{Z}_n$. A formal point of view is that the elements of $\mathbb{Z}_n$ are the cosets of

$n\mathbb{Z}$ in $\mathbb{Z}$. This point of view is interesting, but we will stick to the congruence notation when working with $\mathbb{Z}_n$.

We introduce a standard method of using well-ordering[1] in proofs, a method that is more sophisticated than what we have encountered before this. Quotient groups give us a setting in which to apply this method to obtain results that are interesting in their own right.

The general setting is that we are trying to prove a theorem about finite groups, and we assume, to get a contradiction, that the theorem is false. Let S be the set of natural numbers n such that there is a group with n elements for which the theorem is false. The assumption that the theorem is false amounts to the assumption that S is non-empty. Well-ordering then gives the non-empty set S a minimal element n. By definition of S, there is then a group G with $|G| = n$ such that the theorem is false for G. Furthermore, if K is a group and $|K| < |G|$, then the minimality of $|G|$ in S shows that the theorem holds for K. (Note that K does not have to be a subgroup of G, it only has to have smaller order.) Thus G is a group violating the theorem with as few elements as possible. Such a group is called a *minimal counterexample* to the theorem. We will use the phrase, "let G be a minimal counterexample to the theorem," to presuppose a proof by contradiction under the assumption that the theorem is false, and we will assume you know that the existence of such G is granted by well-ordering.

The use of minimal counterexamples is rampant in algebra. We now illustrate this in the proofs of two interesting and basic facts. We will use quotient groups to produce "groups with less elements." It is this use of quotient groups which we take as a first illustration of their utility.

---

[1] Some people would say we are using induction. Because well-ordering and induction are equivalent, it doesn't really matter.

11. NORMAL SUBGROUPS

LEMMA 11.3. *Let G be an abelian group, let n be a natural number such that $x^n = 1$ for all $x \in G$. Let p be a prime divisor of $|G|$. Then p divides n.*

PROOF. Let G be a minimal counterexample, and choose n and p as in the hypothesis, but with p not dividing n. (Remember that the end of the proof will be to obtain a contradiction! That contradiction will show that G cannot exist, and so the Lemma will hold.)

The existence of p shows that $|G| > 1$, so that there is $y \in G$ with $y \neq 1$. Put $H = \langle y \rangle$, so that

$$(11.3) \qquad 1 < |H|$$

Since G is abelian, $H \triangleleft G$, so that G/H is a group, and it is abelian. Its order $|G/H|$ divides $|G|$ by Lagrange's Theorem, and because of (11.3), this number is less than $|G|$. The minimality of G then forces that the Lemma holds in the group G/H. (Here is the crucial use of the "minimal counterexample.")

Recall the hypothesis for $x \in G$, that $x^n = 1$. We claim that the same hypothesis holds in the group G/H. To show this we need to take an element of G/H and raise it to the n-th power. An element of G/H looks like xH where $x \in G$. Compute that $(xH)^n = (x^n)H$, and since $x^n = 1$, we see that $(xH)^n = x^n H = 1H = H$. (Recall that H is the identity element of G/H.) Thus, the hypothesis about n holds in G/H.

If the prime p were a divisor of $|G/H|$, then the veracity of the Lemma for G/H would force that p divides n, which is not the case.

We conclude that p does not divide $|G/H|$. But p does divide $|G|$, by hypothesis, and since $|G| = |H| \cdot |G/H|$, and because p is a prime, Proposition 8.8 shows that p must divide $|H|$.

Now go back to the definition of H. Recall that $H = \langle y \rangle$ We know that $|H| = o(y)$, so since p divides $|H|$, it must be that p divides $o(y)$. The hypothesis on G includes the fact that $y^n = 1$. By Corollary 7.3, the number $o(y)$

divides n. We have that p divides $o(y)$ and $o(y)$ divides n. We conclude that p divides n, a contradiction. The proof is complete. □

Lemma 11.3 can be used to prove the following.

CAUCHY'S THEOREM FOR ABELIAN GROUPS. *Let G be an abelian group and let p be a prime divisor of $|G|$. Then G has an element of order p.*

PROOF. By the Fundamental Theorem of Arithmetic, we can write $|G| = nq$ where q is a power of p, and such that p does not divide n. Lemma 11.3 then says that there is some $x \in G$ for which $x^n \neq 1$. We will show that p divides $o(x)$. If this is true, then by Proposition 9.3b on the subgroups of cyclic groups, we will know that $\langle x \rangle$ has an element of order p; this will complete the proof.

Corollary 10.2 of Lagrange's Theorem says that $o(x)$ divides $|G| = qn$. Suppose that p does not divide $o(x)$. Since q is a power of p, the numbers q and $o(x)$ have GCD equal to 1. By Proposition 8.5b, $o(x)$ divides n. But then $x^n = 1$, a contradiction. We conclude that p divides $o(x)$. □

Cauchy's Theorem gives the abelian group G a subgroup of order p for every prime p dividing $|G|$. Notice that this is a partial converse to Lagrange's Theorem. We have noted that the converse is not true in general, but it is true for abelian groups, as we now move toward proving. To do this, we will study a minimal counterexample, but we will need to describe the subgroups of a quotient group G/H in terms of H. This task merits its own theorem. Observe that if H is a subgroup of J and J is a subgroup of G, with $H \triangleleft G$, then $H \triangleleft J$ (see Proposition 11.1). The following proof is difficult only in keeping a clear distinction between the two groups G and G/H.

CORRESPONDENCE THEOREM. *Let H be a normal subgroup of the finite group G. Then there is a one to one, onto function from the set of subgroups of G/H to the set of subgroups of G containing H. Specifically,*

(a) *If H is a subgroup of J and J is a subgroup of G, then J/H is a subgroup of G/H;*

(b) *If $\mathbb{J}$ is a subgroup of G/H, let $J = \{x \in G \mid xH \in \mathbb{J}\}$, and then H is a subgroup of J and J is a subgroup of G with $J/H = \mathbb{J}$.*

(c) *Using the notation of (a) and (b), the subgroup $\mathbb{J}$ is normal in G/H if and only if J is normal in G.*

PROOF. For (a), since $H \triangleleft J$, the group J/H is defined. Also, since J/H is a set of left cosets of H, it is a subset of the set G/H of all left cosets of H. Furthermore, the operation on G/H and J/H is the same (coset multiplication). This proves that J/H is a subgroup of G/H.

For (b), the identity element of G/H must be an element of $\mathbb{J}$, and this identity element is H: thus $H \in \mathbb{J}$. If $x \in H$, then $xH = H \in \mathbb{J}$, so that the definition of J yields $x \in J$. We see that J is a subset of G containing H. Given $x, y \in J$, the definition shows that $xH, yH \in \mathbb{J}$. Since $\mathbb{J}$ is a group, we have $xyH = xH \cdot yH \in \mathbb{J}$, and it follows that $xy \in J$. Proposition 9.1 shows that the set J is a subgroup of G. Since $H \subseteq J$, the group H is a subgroup of J.

For $x \in G$, we have that $xH \in \mathbb{J}$ if and only if $x \in J$, by the definition of J, and this shows that $J/H = \mathbb{J}$. Now (b) holds.

Statement (a) shows that the function $f(J) = J/H$ takes subgroups of G containing H to subgroups of G/H, and statement (b) says that f is onto. We complete the proof by showing that f is one to one. Indeed, suppose that $J/H = K/H$, and we will show that $J = K$. Indeed, if $x \in J$, then $xH \in J/H = K/H$. The cosets of H in K are subsets of K, and we see that $x \in K$. We have proved that J is a subgroup of K. The proof of the other

11. NORMAL SUBGROUPS

containment is similar. This proves (b).

You are left to prove (c). □

Now we use the Correspondence Theorem to strengthen Cauchy's Theorem for Abelian Groups.

THEOREM 11.4. *Let G be an abelian group, and let m be a natural number dividing $|G|$. Then G has a subgroup of order m.*

PROOF. Let G be a minimal counterexample, and choose a natural number m dividing $|G|$, where G has no subgroup of order m. Since G does have a subgroup of order 1, we must have $m > 1$, so that m has a prime divisor p.

By Cauchy's Theorem for Abelian Groups, the group G has a subgroup H of order p. That G is abelian yields $H \triangleleft G$, and since $|H| > 1$, we have that $|G/H| < |G|$, so that our present Theorem holds for the group G/H (since G is a minimal counterexample).

To apply the Theorem to G/H we need a divisor of $|G/H|$. Remembering that m divides $|G|$, we can write $|G| = qm$ for some positive integer q. Now also, p divides m, and we write $m = rp$. Thus $|G| = qrp$. On the other hand,

$$qrp = |G| = |G/H| \cdot |H| = |G/H| \cdot p$$

which proves that $qr = |G/H|$. What we want here is that r divides $|G/H|$.

Since the Theorem holds for G/H, this group must have a subgroup $\mathbb{J}$ of order r. By the Correspondence Theorem, the group $\mathbb{J}$ can be written J/H where J is a subgroup of G. We claim that $|J| = m$. Indeed,

$$|J| = |J/H| \cdot |H| = |\mathbb{J}| \cdot |H| = rp = m$$

This gives G a subgroup of order m, a contradiction. □

Under the hypothesis of Theorem 11.4, if G is cyclic, then Proposition 9.3 says that there is only one subgroup of G having order m. Note that K_4 has three subgroups of order 2, so that the subgroups given by Theorem 11.4 do not have to be unique.

Both Theorem 11.4 and Lemma 11.3 have the following curious property: a statement X is assumed false, and then shown to be true; this contradiction proves that X was true in the first place. In other words, X is true, even if it is false! Such a statement must be true, indeed.

In a group G, the subgroup consisting of the identity element is normal, as is the entire group itself. A group with no other normal subgroups is called a *simple group*. A finite abelian simple group has to have order 1 or order p for some prime p, for if the order has a proper divisor m, with $m > 1$, then Theorem 11.4 gives the group a subgroup of order m. Because the group is abelian, the subgroup of order m is normal.

The non-abelian finite simple groups are much harder to find; there are infinitely many of them, although they have been classified in fairly well understood "families," along with a short list of "exceptional groups" that do not fit into the families. The typical family consists of groups of matrices. The alternating groups $\mathbb{A}_n$ are simple when $n \neq 4$. It turns out that all non-abelian simple groups have order divisible by 2; the theorem to that effect, established by Feit and Thompson in the early 1960's via a proof that occupies hundreds of pages, led to a 20-year period of extremely intense research by hundreds of mathematicians, resulting eventually in the solution of the general classification problem. An introduction to this amazing story can be found in *Finite Simple Groups*, by D. Gorenstein (Plenum, 1982).

We establish the simplicity of the alternating groups. The following proof is quite computational; even so, the result is of historical importance, and your ability to follow the calculation is a good test of your understanding of permutations.

THEOREM 11.5. *Let n be a positive integer. If $n \neq 4$, then $\mathbb{A}_n$ is simple.*

11. NORMAL SUBGROUPS

PROOF. The groups $\mathbb{A}_1$ and $\mathbb{A}_2$ have order 1; the group $\mathbb{A}_3$ has order 2, and so these are all simple. Let $n \geq 5$, and suppose that $N \triangleleft \mathbb{A}_n$ with $|N| > 1$, and we will show that $N = \mathbb{A}_n$.

Let $x \in N$ with $x \neq 1$ and chosen with a maximal number of fixed points. There is a prime p dividing the order of x, and if $o(x) = p \cdot q$, then x^q has order p. Furthermore, x^q has no less fixed points than does x, and so x^q has a maximal number of fixed points. We replace x by x^q, and now x has prime order p. We will show that $p = 3$ and that x is a 3-cycle.

Suppose that $p \geq 5$. Writing x as disjoint cycles, there must be a p-cycle involved. Say $x = [abcde\ldots]\cdots$, where a, b, c, d, e are distinct points. Let $y = [ab][cd]$ so that $y \in \mathbb{A}_n$, and $yxy^{-1}x \in N$, since N is normal. Compute that $yxy^{-1}x = [a][be\ldots]\cdots$, so that this non-identity element has more fixed points than does x, a contradiction.

We have $p \leq 3$. Suppose that $p = 2$ and that x consists of two 2-cycles $[ab][cd][e]\cdots$. Let $y = [cde] \in \mathbb{A}_n$ and compute that $yxy^{-1}x = [a][b][ced]\cdots$, a non-identity element with more fixed points than x. Thus, x has at least three 2-cycles: $[ab][cd][ef]\cdots$. Let $y = [cde]$ and compute $yxy^{-1}x = [a][b][ce][df]\cdots$ has more fixed points than x. Thus, $p \neq 2$. We have proved that $p = 3$.

Now suppose that x has at least two 3-cycles: $[abc][def]\cdots$. Let $y = [cde]$, and compute that $yxy^{-1}x = [adcbf][e]$, more fixed points! We conclude that x is a 3-cycle.

The next part of the proof will show that every 3-cycle is in N. First consider a 3-cycle z that has exactly two points in common with x: if $x = [abc]$, then we can imagine that $z = [abd]$ or that $z = [bad]$. In the latter case, define $y = [ab][cd] \in \mathbb{A}_n$ and compute that $yxy^{-1} = z$. In the former case, $yxy^{-1} = z^{-1}$, so that $z^{-1} \in N$, whence $z \in N$ in this case as well.

Since every three cycle can be obtained starting with x and changing one point at a time, every 3-cycle is in N.

A homework problem shows that every element of $\mathbb{A}_n$ is a product of 3-cycles. Since every 3-cycle is in N, we see that $N = \mathbb{A}_n$, and the proof is complete. $\square$

Note. The group $\mathbb{A}_4$ is *not simple*, since

$$\{\, E,\ [12][34],\ [13][24],\ [14][23]\,\} \triangleleft \mathbb{A}_4$$

Problems

137. Every subgroup of the center of a finite group is a normal subgroup of the group.

138. Recall the quaternion group Q_8, described on p.75. Show that $\langle x \rangle$ and $\langle y \rangle$ and $\langle xy \rangle$ and $\langle z \rangle$ are normal subgroups of Q_8. (Note: there are only two other subgroups: Q_8 itself and $\{1\}$. Thus, all subgroups of Q_8 are normal. It is surprising, then, that Q_8 is not abelian.)

139. Let G be a finite group and suppose that G has exactly one element of order 2. Show that this element is in the center of G.

140. If G is a group, and $K \triangleleft G$ and H is a subgroup of G, then $H \cdot K$ is a subgroup of G.

141. Write down the elements and multiplication table for $U_{36}/\langle 5 \rangle$.

142. Suppose that N, K are groups and that $G = N \times K$. Define $\bar{N}$ to be the set of all $(n, 1)$ such that $n \in N$. Prove that $\bar{N} \triangleleft G$.

143. Let G be a finite group and let H be a subgroup of G, with $|G/H| = 2$. Show that $H \triangleleft G$.

144. Prove statement (c) in the Correspondence Theorem.

11. NORMAL SUBGROUPS

145. Let G be a finite group and let H be a subgroup of G. Define
$$N(H) = \{x \in G \mid y \in H \Rightarrow x \cdot y \cdot x^{-1} \in H\}$$
Show that $N(H)$ is a subgroup of G and that $H \triangleleft N(H)$. (Note: the group $N(H)$ is the *normalizer of H in G*.)

146. Show that $\langle [13] \rangle \triangleleft \langle [13], [24] \rangle$ and $\langle [13], [24] \rangle \triangleleft D_8$, but $\langle [13] \rangle \not\triangleleft D_8$.

147. Let G be an abelian group, and write $|G| = q \cdot n$ where q, n are positive integers with GCD equal to 1. By Theorem 11.4, G has a subgroup H of order q. Let $x \in G$ with $x^q = 1$. Show that $x \in H$. (Hint: what is the order of xH in G/H?)

148. Notice that $|\mathbb{S}_3| = 2 \cdot 3$. Define H to be the set of $x \in \mathbb{S}_3$ such that $x^2 = E$. Show that H is not a subgroup of $\mathbb{S}_3$.

149. Let p be an odd prime. Show that there are no simple groups of order $4 \cdot p$. Let G be a group of order $4 \cdot p$, and you might want to follow this outline:

(a) A previous problem shows that G has an element of order 2. If it has only one element of order 2, it is not simple.

(b) Let x, y be distinct elements of order 2, and $H = \langle x, y \rangle$. Then $|H|$ is even, greater than 2, and divides $4 \cdot p$.

(c) If $|H| = 4 \cdot p$, then G is not simple. (Recall the structure of H.)

(d) If $|H| = 2 \cdot p$, then G is not simple. (Previous problem: $|G/H| = 2$.)

(e) We can assume that $|H| = 4$. Then x, y commute. Thus, we can assume that the elements of G of order 2 all commute.

(f) Let K be the set of all elements of order 2, along with 1_G. Show that K is an abelian subgroup of G, and $K \triangleleft G$. Thus, $K = G$.

(g) By Theorem 11.4, G is not simple.

CHAPTER 12

Group Homomorphisms and Isomorphisms

For groups G, K we want to call attention to their operations; write $x \circ y$ for the operation of G and $a \bullet b$ for the operation of K. A function $f : G \to K$ is a *group homomorphism* if we have

(12.1) $$f(x \circ y) = f(x) \bullet f(y) \qquad \text{for all} \quad x, y \in G$$

Note that the operation of G is inside f and that of K is outside f. When we identify a homomorphism, we will usually revert to the practice of not calling attention to the operations involved.

An important example is given by the parity function of a permutation. Statement (d) of the Parity Theorem shows that $\sigma : \mathbb{S}_P \to \mathbb{Z}_2$ satisfies (12.1), and so σ is a group homomorphism.

Here is a less interesting example. If G is an abelian group and n is an integer, then $f : G \to G$ defined by $f(x) = x^n$ is a homomorphism. Note that if $G = D_8$ (which is non-abelian), the square function $f(x) = x^2$ is **not** a homomorphism.

The paradigm example of a group homomorphism occurs when we have a normal subgroup H of a group G. Define $f : G \to G/H$ by

$$f(x) = xH$$

That f is a group homomorphism follows directly from the definition of the group operation on G/H (recall Theorem 11.2). Here is the calculation:

$$f(xy) = xyH = xH \cdot yH = f(x) \cdot f(y) \quad \text{for all} \quad x, y \in G$$

We call f the *canonical homomorphism* from G onto G/H. The function f is determined by the normal subgroup H. Let us show that the normal subgroup is determined by the function If $x \in H$, then $f(x) = xH = H$, so that x maps to the identity element of the quotient group G/H. Conversely, if $f(x) = H$, then $xH = H$, whence $x \in H$. Thus, H is the set of $x \in G$ such that $f(x) = 1$.

Taking this idea back to an arbitrary group homomorphism $f : G \to K$ we define the *kernel* $\ker(f)$ of f

$$\ker(f) = \{x \in G \mid f(x) = 1\}$$

We will need the following facts about group homomorphisms, one of which is that $\ker(f)$ is a normal subgroup of G.

PROPOSITION 12.1. *Let $f : G \to K$ be a group homomorphism, where G is a finite group.*

(a) *We have $f(1) = 1$ and $f(x^{-1}) = f(x)^{-1}$ for all $x \in G$.*
(b) *$f(G)$ is a subgroup of K.*
(c) *$\ker(f)$ is a normal subgroup of G.*
(d) *For $x, y \in G$ we have $f(x) = f(y)$ if and only if $x \cdot \ker(f) = y \cdot \ker(f)$, so that the function f is one to one if and only if $\ker(f) = \{1\}$.*
(e) *$|f(G)| = |G/\ker(f)|$.*
(f) *If $x \in G$, then the order of $f(x)$ divides the order of x.*

PROOF. Compute $f(1) = f(1 \cdot 1) = f(1) \cdot f(1)$. Multiplying both sides of $f(1) = f(1)f(1)$ by $f(1)^{-1}$ we obtain $f(1) = 1$. Then for $x \in G$, we have $1 = f(1) = f(xx^{-1}) = f(x)f(x^{-1})$, so that $f(x^{-1})$ is the inverse of $f(x)$. We have proved (a).

You can prove that $f(G)$ and $\ker(f)$ are subgroups. Given $y \in \ker(f)$ and $x \in G$, we have

$$f(x^{-1}yx) = f(x)^{-1}f(y)f(x)$$
$$= f(x)^{-1} \cdot 1 \cdot f(x)$$
$$= f(x)^{-1}f(x) = 1$$

and this shows that $x^{-1}yx \in \ker(f)$. By the definition of normal subgroup (given after Proposition 11.1), we have $\ker(f) \triangleleft G$, and (c) is proved.

For (d), if $f(x) = f(y)$, then use (a) to see that

$$f(x^{-1} \cdot y) = f(x)^{-1} \cdot f(y) = 1$$

We see that $x^{-1} \cdot y \in \ker(f)$, so that $x \cdot \ker(f) = y \cdot \ker(f)$ by Proposition 10.4. By the same proposition, if $x \cdot \ker(f) = y \cdot \ker(f)$, then $x = y \cdot z$ where $z \in \ker(f)$, and we have

$$f(x) = f(y \cdot z) = f(y) \cdot f(z) = f(y)$$

This proves the if and only if statement in (d). We leave the statement about f being one to one to you.

Statements (e) and (f) are left to you. □

We mentioned that the parity function $\sigma : \mathbb{S}_P \to \mathbb{Z}_2$ is a homomorphism. Notice that $\mathbb{A}_P$ is the kernel of this homomorphism. We already know that $\mathbb{A}_P \triangleleft \mathbb{S}_P$; if you go back to the proof in Chapter 11, you will see that it is a special case of the general proof of Proposition 12.1c.

Before stating Proposition 12.1, we considered the situation of $H \triangleleft G$ and we defined the canonical homomorphism from G onto G/H. The kernel of this homomorphism is H. In other words, every normal subgroup is the kernel of a homomorphism. Proposition 12.1 shows that every kernel is a normal subgroup. This gives an alternative way to introduce and study normal subgroups. Instead of defining them as we did, and then defining the homomorphism, you could define the homomorphism first, and then define the normal subgroup as the kernel. The approach appeals to those who find the homomorphism to be "natural" from the point of view of abstract group theory. Our approach is more historical and allows us to work more natural examples as we go along.

Here is a very important way to manufacture homomorphisms. Let G be a group and H a subgroup of G (do not assume that H is normal). For $x \in G$, we define a permutation $f(x)$ of G/H. (We repeat: H is not necessarily normal, and so G/H is just the set of left cosets of H in G.) To define $f(x)$ as a permutation on G/H, we must describe it as a function on the left cosets of H in G. We define

$$f(x)(yH) = xyH \quad \text{for all} \quad y \in G$$

We call f the *coset homomorphism* for H in G. It is very helpful to compute f explicitly in some example! We suggest $G = D_8$, and have H be a cyclic subgroup of order 2.

The following result is not really hard; but it is hard to keep track of its many different objects. Read it slowly.

PROPOSITION 12.2. *Let H be a subgroup of the group G, and let f be the coset homomorphism for H in G. Then f is a homomorphism from G into the group of permutations on the cosets G/H. The kernel of f is a subgroup of H.*

PROOF. Let $x \in G$, and we will show that $f(x)$ is a permutation on G/H. Indeed, $f(x)$ is onto since if $y \in G$, then

$$f(x)(x^{-1}yH) = xx^{-1}yH = yH$$

so that yH is in the image of $f(x)$.

The function $f(x)$ is one to one, since

$$f(x)(yH) = f(x)(zH) \quad \Rightarrow \quad xyH = xzH$$

This implies that $yH = zH$. Thus, $f(x)$ is a permutation.

Next we show that f is a homomorphism. If $x, y \in G$, then we need to show that
$$f(x)f(y) = f(xy)$$
The objects on each side of this equation are functions (permutations), and so we need to show that the two sides agree as functions. Thus, for $z \in G$, we need to show that
$$f(x)\left(f(y)(zH)\right) = f(xy)(zH)$$
The left side is
$$f(x)(yzH) = xyzH$$
and this is the right side as well.

Finally, let $k \in \ker(f)$. Then $f(k)$ is the identity permutation. In particular, $f(k)(H) = H$, so that $kH = H$. This implies that $k \in H$, as needed. □

Here is a useful divisibility fact related to the coset homomorphism.

COROLLARY 12.3. *In the context of Proposition 12.2, we have that* $|G/\ker(f)|$ *divides* $|G/H|!$.

PROOF. Let S be the group of permutations on the cosets G/H. Let $n = |G/H|$, so that $|S| = n!$.

Proposition 12.1(b) shows that $f(G)$ is a subgroup of S. By Lagrange's Theorem, $|f(G)|$ divides $|S| = n!$. Proposition 12.1(e) says that $|f(G)| = |G/\ker(f)|$, and we are done. □

Another application: a problem in Chapter 11 asked you to show that if H is a subgroup of G with $|G/H| = 2$, then $H \triangleleft G$. Here is a generilzation: if p is the smallest prime dividing $|G|$, if H is a subgroup of G, and if $|G/H| = p$, then $H \triangleleft G$. Indeed, let $f : G \to \mathbb{S}_p$ be the coset homomorphism, and Corollary 12.3 says that $|G/\ker(f)|$ divides $p!$. Since p is the smallest prime dividing $|G|$, we see that $(p-1)!$ has GCD 1 with $|G|$. Thus, $|G/\ker(f)|$ divides

p. Proposition 12.2 says that $\ker(f) \subseteq H$, and this proves that $|G/\ker(f)| > 1$. Thus, $|G/\ker(f)| = p$, and it follows that $\ker(f) = H$. Therefore, $H \triangleleft G$.

Here is an important application of the application!

PROPOSITION 12.4. *Let G be a group of order p^2 where p is prime. Then G is abelian.*

PROOF. Let $x, y \in G$, and we will show that $xy = yx$. By Corollary 10.2, each of x, y has order 1 or p or p^2.

If either x or y has order p^2, then G is cyclic, hence abelian, and we are done. If either of them has order 1, then $xy = yx$, by the properties of the identity element. It remains to assume that x and y have order p. If $y \in \langle x \rangle$, then $xy = yx$, and so we can assume that $y \notin \langle x \rangle$.

Because $y \notin \langle x \rangle$, the subgroup $\langle x \rangle \cap \langle y \rangle$ is proper in $\langle x \rangle$, a subgroup of prime order. Thus, only the identity element can be in $\langle x \rangle \cap \langle y \rangle$

We have $|G/\langle x \rangle| = p$, and so, by the problem above, we have $\langle x \rangle \triangleleft G$. Similarly, $\langle y \rangle \triangleleft G$. Now consider the element $xyx^{-1}y^{-1}$. This can be written

$$xyx^{-1}y^{-1} = (xyx^{-1})y^{-1} = x(yx^{-1}y^{-1})$$

In light of Proposition 11.1, this element is in both normal subgroups $\langle x \rangle$ and $\langle y \rangle$, and their intersection is the identity element. Thus, $xyx^{-1}y^{-1} = 1$. It follows that $xy = yx$. □

A *group isomorphism* is a homomorphism that is one to one and onto. By Proposition 12.1d, the kernel of an isomorphism consists only of the identity element (of the domain of the function). If there is a group isomorphism $f : G \to K$, then we say that G and K are *isomorphic*. Intuitively, this means that G and K are the *same group*. Indeed, given $h, k \in K$, that f is one to one and onto yields unique elements x and y of G such that $f(x) = h$ and $f(y) = k$. Because f is a homomorphism, $hk = f(x)f(y) = f(xy)$, so that the product hk in K is determined by the function f and the product in G. Thus the isomorphism f is a pairing between the two groups G and K, in which the group operations of the two groups correspond.

If $f : G \to K$ is an isomorphism, then f^{-1} is an isomorphism as well, as you can check. If $f : G \to K$ is an isomorphism, and if $g : K \to H$ is an isomorphism, then $gf : G \to H$ is an isomorphism, too. Thus, groups isomorphic to the same group are isomorphic to each other.

We write $G \cong K$. when the groups G, K are isomorphic. The discussion of the two previous paragraphs shows that isomorphism is an equivalence relation.

Every group of prime order p is isomorphic to $\mathbb{Z}_p$ (a cyclic group of order p). Every group of order 4 is isomorphic to $\mathbb{Z}_4$ or to K_4 and not to both. Every group of order 6 is isomorphic to $\mathbb{Z}_6$ or to D_6; for example, $D_6 \cong \mathbb{S}_3$. The groups D_8 and Q_8 are non-isomorphic groups of order 8.

One way to describe group theory abstractly is as the study of properties of groups that are preserved under isomorphisms. Here are some examples of such properties (there are many more); you may find yourself providing some or all of the proofs. Throughout, suppose that $f : G \to K$ is an isomorphism.

(a) We have $|G| = |K|$.
(b) G is abelian if and only if K is abelian.
(c) If $x \in G$, then $o(x) = o(f(x))$. Thus, the groups G, K have the same numbers of elements of the same orders.
(d) If H is a subgroup of G, then $f(H)$ is a subgroup of K that is isomorphic to H. Every subgroup of K comes about as $f(H)$ for a unique subgroup H of G. We have $H \triangleleft G$ if and only if $f(H) \triangleleft K$. In particular, G, K have the same numbers of subgroups and the same numbers of normal subgroups.

There is an isomorphism associated with every homomorphism, and this association gives a further link between quotient groups and homomorphisms.

Let $f : G \to K$ be a group homomorphism. As we have seen in Proposition 12.1, if $x \in G$, then f maps the entire coset $x \cdot \ker(f)$ to the same element $f(x)$ of K. We can therefore define a function $F : G/\ker(f) \to f(G)$ by $F(x \cdot \ker(f)) = f(x)$ for all $x \in G$.

PROPOSITION 12.5. *Given the group homomorphism $f : G \to K$, the function $F : G/ker(f) \to f(G)$ defined by $F(x \cdot \ker(f)) = f(x)$ is a group isomorphism.*

PROOF. Compute for $x, y \in G$ that
$$F(x \cdot \ker(f) \cdot y \cdot \ker(f)) = F(xy \cdot \ker(f)) = f(xy)$$
$$= f(x)f(y) = F(x \cdot \ker(f))F(y \cdot \ker(f))$$
so that F is indeed a group homomorphism. We use Proposition 12.1d to show that F is one to one. Let $x \cdot \ker(f) \in \ker(F)$, then by definition
$$1 = F(x \cdot \ker(f)) = f(x)$$
so that $x \in \ker(f)$. But then $x \cdot \ker(f) = \ker(f)$, and this last coset is the identity element of the quotient group $G/\ker(f)$. Thus F is one to one.

To show that F is onto, each $k \in f(G)$ has the form $f(x)$ for some $x \in G$. Then $F(x \cdot \ker(f)) = k$. □

Proposition 12.5 shows that the homomorphic images of a group G are quotient groups of G. In other words, the internal structure of G (including its normal subgroups and quotients) accounts for its homomorphisms. This proposition should be taken with the Correspondence Theorem. If $f : G \to K$ is a homomorphism, then the subgroups of $f(G)$ are in one to one correspondence with those subgroups of G which contain $\ker(f)$.

Here is a specific example, which ought to be expected.

PROPOSITION 12.6. *If G is a cyclic group of order n, then G is isomorphic to $\mathbb{Z}_n$.*

PROOF. Remember the formality: $\mathbb{Z}_n = \mathbb{Z}/(n\mathbb{Z})$.

Let $G = \langle x \rangle$, and consider the function $f : \mathbb{Z} \to G$ defined by $f(m) = x^m$. We have remarked that this is a homomorphism, since G is abelian. The definition of $\langle x \rangle$ shows that f is onto. The kernel of f is the set of integers m such that $x^m = 1$. By Corollary 7.3, $x^m = 1$ if and only if m is divisible by n. In other words, the kernel is $n\mathbb{Z}$. By Proposition 12.5, we have $\mathbb{Z}/(n\mathbb{Z})$ isomorphic to $\langle x \rangle = G$. $\square$

Next we indulge in a level of generality that may strain you a little bit. Let G be a finite group. An isomorphism $f : G \to G$ is called an *automorphism*. The set of automorphisms is denoted $\mathrm{Aut}(G)$, and it is a group under function composition. This is straightforward, and you should verify it yourself.

Here is an example.

PROPOSITION 12.7. *Let G be a cyclic group of order n. Then $\mathrm{Aut}(G)$ is isomorphic to U_n. In particular, $\mathrm{Aut}(\mathbb{Z}_n)$ is isomorphic to U_n.*

PROOF. Let $G = \langle x \rangle$. If $f \in \mathrm{Aut}(G)$, then $f(x) \in G$, and so $f(x) = x^k$ for some integer k. Because f is an isomorphism, the elements x and $x^k = f(x)$ have the same order. By Corollary 8.4, we have that $k \in U_n$. We write $k = I(f)$. The function I maps $\mathrm{Aut}(G)$ to U_n. We will show that I is an isomorphism.

If $f, g \in \mathrm{Aut}(G)$, compute that

$$x^{I(f \cdot g)} = f \cdot g(x) = f(g(x)) = f(x^{I(g)}) = f(x)^{I(g)} = (x^{I(f)})^{I(g)} = x^{I(f)I(g)}$$

This shows that $I(f \cdot g) \equiv I(f)I(g) \bmod n$. In other words, I is a homomorphism.

We will show that I is one to one and onto. If $f, g \in \text{Aut}(G)$ with $I(f) = I(g)$, then we claim that $f = g$. To show this, we need to show that f and g are the same function on G. Elements of G are powers of x:

$$f(x^k) = f(x)^k = x^{I(f)k} = x^{I(g)k} = g(x)^k = g(x^k)$$

We see that $f = g$, so that I is one to one.

Finally, we show that I is onto. Let $k \in U_n$, and define $f : G \to G$ by $f(y) = y^k$. We leave it to you to show that f is an automorphism of G (in other words, f is a homomorphism, one to one, and onto). □

Here is another way that isomorphisms come about naturally. The proof is left to you.

PROPOSITION 12.8. *Let G be a group, and $x \in G$, and H a subgroup of G. Define $f(x) : H \to G$ by $f(x)(h) = xhx^{-1}$. Then $f(x)$ is an isomorphism from H to the subgroup xHx^{-1}. In particular, if $H \triangleleft G$, then $f(x)$ is an automorphism of H.*

Proposition 12.8 gives us a general way to construct what might be called *compound groups*. To explain how this works, suppose that $N \triangleleft G$ and let H be a subgroup of G. For each $h \in H$, Proposition 12.8 says that the mapping $f : N \to N$ defined by $f(x) = h \cdot x \cdot h^{-1}$ is an automorphism of N.

You have proved that $N \cdot H$ is a group; we want to show that multiplication in this group is determined by

(a) multiplication in N;
(b) multiplication in H;
(c) $h \cdot y \cdot h^{-1}$ for $h \in H$ and $y \in N$.

Here is what we mean: two elements of NH look like this: $x \cdot h$ for $x \in N$ and $h \in H$, and $y \cdot k$ for $y \in N$ and $k \in H$. Observe that

$$(x \cdot h) \cdot (y \cdot k) = x \cdot h \cdot y \cdot k = x \cdot h \cdot y \cdot h^{-1} \cdot h \cdot k = x \cdot (h \cdot y \cdot h^{-1}) \cdot h \cdot k$$

In the final term, the rightmost product $h \cdot k$ is determined by multiplication in H; the middle term $h \cdot y \cdot h^{-1}$ is of the form identified in (c); the leftmost product occurs in N.

To repeat: multiplication in NH is determined by multiplication in N and H and by the terms (c), which Proposition 12.8 says come from automorphisms of N. We want to turn this around backwards: given a group N, and given a subgroup H of $\text{Aut}(N)$, we show how to *construct* a group that will have N and H as subgroups, with N normal. In the following, remember that if H is a subgroup of $\text{Aut}(N)$, then the elements of H are *functions* on N.

PROPOSITION 12.9. *Let N be a group and let H be a subgroup of $\text{Aut}(N)$. Define an operation on $N \times H$ by the formula $(x, h) \cdot (y, k) = (x \cdot h(y), h \cdot k)$. Then*

(a) *$N \times H$ is a group under this operation;*
(b) *if N' is the set of $(x, 1)$ for all $x \in N$, then N' is a normal subgroup of $N \times H$, and $N' \cong N$;*
(c) *if H' is the set of $(1, h)$ for all $h \in H$, then H' is a subgroup of $N \times H$, and $H' \cong H$;*
(d) *we have $(1, h) \cdot (x, 1) \cdot (1, h)^{-1} = (h(x), 1)$.*

PROOF. The elements of H are functions and their operation is function composition. Also, the elements of H are homomorphisms. The associative law is tedious but straightforward:

$$\left[(x, h) \cdot (y, k)\right] \cdot (z, m)$$
$$= (x \cdot h(y), h \cdot k) \cdot (z, m) = (x \cdot h(y) \cdot (hk)(z), h \cdot k \cdot m)$$
$$= (x \cdot h(y) \cdot h(k(z)), h \cdot k \cdot m) = (x \cdot h(y \cdot k(z)), h \cdot k \cdot m)$$
$$= (x, h) \cdot (y \cdot k(z), k \cdot m) = (x, y) \cdot \left[(y, k) \cdot (z, m)\right]$$

It is easy to check that the identity element is $(1,1)$ and that the inverse of (x, h) is $(h^{-1}(x^{-1}), h^{-1})$. Thus, $N \times H$ is a group.

We leave the other assertions to you. $\square$

Note that the group $N \times H$ just constructed is **not** the direct product, since the multiplication honors the automorphism action of the elements of H.

Here is an example. Let $\langle x \rangle$ have order 4. Proposition 12.7 gives us the automorphism group of $\langle x \rangle$; let k be the automorphism where $k(x) = x^{-1}$ (using that $-1 \in U_4$). Compute that $k^2(x) = k(k(x)) = k(x^{-1}) = x$, so that k^2 is the identity automorphism. The group $\langle x \rangle \times \langle k \rangle$ defined as in Proposition 12.9 has order 8 with a normal subgroup of order 4 and each of the 4 elements not in this subgroup has order 2 (compute!). Look familiar? In fact, this group is isomorphic to D_8. Many examples of groups can be constructed this way – it is one of the most common construction techniques. The group $N \times H$ of Proposition 12.9 is called a *semi-direct product*.

In Chapter 11, we found normal subgroups of the direct product $N \times H$: the set N' of $(n, 1)$ for all $n \in N$, and the set H' of $(1, h)$ for all $h \in H$. Here is a situation where such a group arises.

PROPOSITION 12.10. *Let N and K be normal subgroups of the group G and suppose that $N \cap K$ consists only of the identity element. Then NK is a subgroup of G that is isomorphic to the direct product $N \times K$.*

PROOF. Define $f : N \times K \to G$ by $f(n, k) = nk$, where the operation on the right is that of G. We leave it to you to show that f is one to one, onto its image NK. You will need the fact that $nk = kn$ for all $n \in N$ and $k \in K$. (Be careful with this: we are not saying that N or K is abelian, only that their elements commute with each other.) The identity $nk = kn$ follows from considering

$$nkn^{-1}k^{-1} = (nkn^{-1})k^{-1} = n(kn^{-1}k^{-1})$$

We did a similar cartwheel in the proof of Proposition 12.4. □

Let A, B, C be groups. We can form the direct product $A \times B$ and then $(A \times B) \times C$. We can also form the direct product $A \times (B \times C)$. As you probably expect, we can think of $(A \times B) \times C$ and $A \times (B \times C)$ as the same – as the set of *ordered triples* (a, b, c) where $a \in A$ and $b \in B$ and $c \in C$. In general, we can have direct products of an arbitrary set of finite groups. The generalization of Proposition 12.10 is not hard to see. We need the mapping $f(a, b, c) = abc$ to be one to one and onto in order for the image of f to be isomorphic with $A \times B \times C$.

We can put Proposition 12.10 together with Theorem 11.4 to deduce the general structure of a finite abelian group. This result is considered basic, and it has applications. We don't have time to go further than to give a proof. And to apologize for the fact that the proof is complicated in an unpleasant way.

FUNDAMENTAL THEOREM OF ABELIAN GROUPS. *Let G be a finite abelian group, and write $|G| = \prod_{i=1}^{m} q_i$ where each q_i is a power of a distinct prime. Then G is isomorphic to a direct product of abelian groups of order q_i, each of which is isomorphic to a direct product of cyclic groups. Thus, a finite abelian group is isomorphic to a direct product of cyclic groups.*

PROOF. By Theorem 11.4, G has subgroups Q and R of order q_1 of order $q_2 \cdots q_m$, respectively. Each of these subgroups is normal, since G is abelian. Lagrange's Theorem shows that $|Q \cap R|$ divides both $|Q| = q_1$ and $|R|$. These latter two numbers have GCD equal to 1, and so $|Q \cap R| = 1$. It follows from Proposition 10.5 that $QR = G$. Proposition 12.10 then shows that G is isomorphic to $Q \times R$. Applying this argument repeatedly (using induction) we see that G is isomorphic to the direct product of groups of order q_i. That wasn't so bad.

Now assume that the order of the abelian group G is a power of the prime p. We will show that G is isomorphic to the direct product of cyclic groups. This is where the argument gets messy.

We claim there is $w \in G$ and a subgroup L of G such that $\langle w \rangle \cap L = \{1\}$ and that $\langle w \rangle L = G$. It follows that G is isomorphic to $\langle w \rangle \times L$. Once this is established, the same fact can be applied to L, that it will be isomorphic to some direct product $L = \langle u \rangle \times M$. We apply the result to M, and so on, and eventually get the original group G isomorphic to a direct product of cyclic groups.

We have reduced the proof to the claim that w and L exist. To prove that they do, assume that G is a minimal counterexample. Since G is finite, it has an element x of maximal order. Then $\langle x \rangle$ is a normal subgroup of G, and $G/\langle x \rangle$ is abelian. By the minimality of G, there is an element Y of $G/\langle x \rangle$, and a subgroup $\mathbb{M}$ of $G/\langle x \rangle$ such that $\langle Y \rangle \cap \mathbb{M} = \{\langle x \rangle\}$ (regarding $\langle x \rangle$ as the identity element of $G/\langle x \rangle$). Also we have $\langle Y \rangle \mathbb{M} = G/\langle x \rangle$.

The definition of $G/\langle x \rangle$ shows that $Y = y\langle x \rangle$ for some $y \in G$. The Correspondence Theorem shows that $\mathbb{M}$ is equal to $L/\langle x \rangle$ where $\langle x \rangle$ is a subgroup of L and L is a subgroup of G. Because $\langle Y \rangle \cap \mathbb{M} = \{\langle x \rangle\}$, we have $\langle y \rangle \cap L$ is a subgroup of $\langle x \rangle$. Because $\langle Y \rangle \mathbb{M} = G/\langle x \rangle$, we have that $\langle y \rangle L = G$. We will use these facts along with the fact that x has maximal order in G.

Let k be the order of $y\langle x \rangle$ in the quotient group (and then k is a power of p). The definition of the order of an element shows that k is the smallest positive integer such that $y^k \in \langle x \rangle$, and since $y^{o(y)} = 1 \in \langle x \rangle$, we see that k divides $o(y)$. Proposition 8.3 then shows that

(12.2) $$k \cdot o(y^k) = o(y)$$

Write $y^k = x^j$ for some integer j, and let q be the GCD of j and $o(x)$, and Proposition 8.3 says that

(12.3) $$q \cdot o(x^j) = o(x)$$

Here is the key calculation; it shows that k divides j. Indeed, remember that x has maximal order in G, that $y^k = x^j$, and use (12.2) and (12.3) to go:

$$q \cdot o(x) \geq q \cdot o(y) = qk \cdot o(y^k) = qk \cdot o(x^j) = k \cdot o(x)$$

We see that $q \geq k$. Since q and k are powers of p, this implies that k divides q. Since q divides j, we see that k divides j. Write $j = ks$.

We claim that G is isomorphic to $\langle yx^{-s}\rangle \times L$. This will be a contradiction that proves the result. Proving that G is isomorphic to the stated direct product will follow from Proposition 12.10 once we show that $G = \langle yx^{-s}\rangle L$ and that $\langle yx^{-s}\rangle \cap L = \{1\}$.

Let $g \in G$. We see that $g = y^a z$ for some $z \in L$. Then $g = y^a z = (yx^{-s})^a x^{sa} z$, and since $\langle x \rangle$ is a subgroup of L, we have that $g \in \langle yx^{-s}\rangle L$. This proves that $G = \langle yx^{-s}\rangle L$.

An element of $\langle yx^{-s}\rangle \cap L$ looks like $(yx^{-s})^t$ for an integer t, and we would have $(yx^{-s})^t \in L$. Then $y^t \in \langle y \rangle \cap L$ is a subgroup of $\langle x \rangle$. Another way to say this is that $y\langle x\rangle$ raised to the t-th power is $\langle x \rangle$. Applying Corollary 7.3 to the quotient group $G/\langle x\rangle$, we see that the order of $y\langle x\rangle$ in the quotient group must divide t. The order of this coset is k, and so k divides t. Write $t = ku$ for an integer u, and then $(yx^{-s})^t = (y^k x^{-sk})^u = (y^k x^{-j})^u = 1$. We have shown that $\langle yx^{-s}\rangle \cap L$ consists only of the identity element. □

The converse of the Fundamental Theorem of Abelian Groups is obvious: a direct product of cyclic groups is abelian.

Problems

150. Suppose that $f : G \to K$ is a group homomorphism. Show that $f(G)$ is a subgroup of K and that $\ker(f)$ is a subgroup of G.

151. Write down the cosets of $\langle [13] \rangle$ in D_8, and label them a, b, c, d (however you wish, but use those letters). For each element x of D_8, find the permutation of a, b, c, d corresponding to x as described before Proposition 12.2 as x permutes the cosets. What is the kernel of this homomorphism?

152. Let P, Q be finite sets, and suppose that $f : P \to Q$ is one to one and onto. For $x \in \mathbb{S}_P$, define $G(x) = f \cdot x \cdot f^{-1}$. Show that $G : \mathbb{S}_P \to \mathbb{S}_Q$ is a group isomorphism.

153. Let G and H be isomorphic groups. Suppose that m, n are integers and G has exactly m elements of order n. Show that H has exactly m elements of order n.

154. Prove Proposition 12.9b and Proposition 12.9d.

155. Prove Proposition 12.8.

156. Let G be a group and $x \in G$. Recall the normalizer $N(\langle x \rangle)$: the set of $y \in G$ such that $y \cdot \langle x \rangle \cdot y^{-1} = \langle x \rangle$. Proposition 12.8 shows that the mapping $f : N \to \text{Aut}\langle x \rangle$, defined by $f(y)(x^k) = y \cdot x^k \cdot y^{-1}$ for all $y \in N$ and $k \in \mathbb{Z}$, is a group homomorphism. Show that $\ker(f) = C_G(x)$. Show that $|N/C_G(x)|$ divides $|U_n|$. (This fact is called the *N/C Theorem*.)

157. The previous problem leads to a shorter proof of Proposition 12.4. We are given $x, y \in G$, and we want to show that $xy = yx$. We can assume that each has order p. The homomorphism argument shows that $\langle x \rangle \triangleleft G$. Use the previous problem to show that $|G/C_G(x)|$ divides $p - 1$. Conclude that $xy = yx$.

CHAPTER 13

Conjugacy Classes

We have shown that the converse of Lagrange's Theorem is false in general but that the converse is true for abelian groups. Sylow's Theorem, often categorized as three separate theorems, deals with the existence of subgroups of prime power order in arbitrary groups. We will prove the existence part of Sylow's Theorem which constructs a subgroup of prime power order whenever the prime power divides the order of the group. Our method of proof will involve a concept that is important in its own right: the *conjugacy class*.

Let G be a group and let $x \in G$. The *conjugacy class of x* is denoted $\text{cl}(x)$; it is a set of elements of G:

$$\text{cl}(x) = \{y \cdot x \cdot y^{-1} \mid y \in G\}$$

To say it in words: for each $y \in G$, we form the product yxy^{-1}; the collection of all these products is the conjugacy class of x. Because we look at each $y \in G$ to get the conjugacy class, you might think that the conjugacy class is a large set – actually its size usually varies from element to element quite a bit, and we will need to work examples to get some intuition.

Let's start with an abelian group G. For $x, y \in G$, compute that $yxy^{-1} = yy^{-1}x = x$. Thus, $\text{cl}(x) = \{x\}$ for each x! Repeat: in an abelian group, each conjugacy class consists of exactly one element.

This prompts a related calculation: notice that $\text{cl}(1) = \{1\}$, since the identity element 1 commutes with every $y \in G$. In fact, we see that if $x \in Z(G)$

(the center), then $\text{cl}(x) = \{x\}$, since x commutes with every $y \in G$. We will return to this fact later.

Here are the conjugacy classes of the elements of $\mathbb{S}_3$. Obviously, you should check this!

$$\text{cl}(E) = \{E\}$$
$$\text{cl}([12]) = \text{cl}([13]) = \text{cl}([23]) = \{[12], [13], [23]\}$$
$$\text{cl}([123]) = \text{cl}([132]) = \{[123], [132]\}$$

There are many suspicious properties of these sets. Let's see what holds in general.

PROPOSITION 13.1. *Let G be a group.*

(a) If $x \in G$, then $x \in \text{cl}(x)$.
(b) If $x, y \in G$ and $x \in \text{cl}(y)$, then $y \in \text{cl}(x)$.
(c) If $x, y, z \in G$ and $x \in \text{cl}(y)$ and $y \in \text{cl}(z)$, then $x \in \text{cl}(z)$.

PROOF. Observe that $x = 1x1^{-1}$, and so $x \in \text{cl}(x)$, proving (a). For (b), let $x \in \text{cl}(y)$, so that $x = wyw^{-1}$ for some $w \in G$. The w-terms can be moved to the left: $w^{-1}xw = y$, and notice that this can be written

$$y = w^{-1}xw = w^{-1} \cdot x \cdot \left(w^{-1}\right)^{-1}$$

In this form, we see that $y \in \text{cl}(x)$, as needed.

To prove (c), let $x \in \text{cl}(y)$ and $y \in \text{cl}(z)$, so that $x = uyu^{-1}$ and $y = vzv^{-1}$. Compute

$$x = uyu^{-1} = uvzv^{-1}u^{-1} = (uv) \cdot z \cdot (uv)^{-1}$$

and this shows that $x \in \text{cl}(z)$. □

A more sophisticated way to give the results of Proposition 13.1: that "x is in the conjugacy class of y" is an equivalence relation.

We computed an example in a permutation group above. There is an easy way to compute yxy^{-1} when x and y are permutations on a finite set A. We will illustrate this, and leave it to you to generalize. Let $x = [12][4][567]$, and let $y = [134][26]$. We write down x, and form yxy^{-1} directly below x, making a copy of the cycle structure of x, and applying y to the points in x to fill in the cycles of yxy^{-1}. Like so

$$x = [12][4][567]$$
$$yxy^{-1} = [36][1][527]$$

We write 3 directly below 1 since y sends 1 to 3. We write 6 below 2 since y sends 2 to 6. In other words, yxy^{-1} sends 3 to 6 and 6 to 3. What we are saying is the if x sends point i to point j, then yxy^{-1} sends $y(i)$ to $y(j)$. You should verify that this is correct and that it accounts for the computation of yxy^{-1}.

Using the above, we can show that $x, w \in \mathbb{S}_A$ are conjugate if and only if they have the same cycle structure. This is important.

Could cl(x) be a subgroup of G?

We observed above that if $x \in Z(G)$, then cl(x) = $\{x\}$. Here is a general counting formula for conjugacy classes; it involves the *centralizer* $C_G(x)$ of a group element x, defined on p.91.

PROPOSITION 13.2. *Let G be a group, and let $x \in G$. Then $|\text{cl}(x)| = |G/C_G(x)|$.*

PROOF. Consider the elements of a left coset $uC_G(x)$, for $u \in G$. These elements look like uc where $c \in C_G(x)$. Use the definition of $C_G(x)$ to compute

$$(uc)x(uc)^{-1} = ucxc^{-1}u^{-1} = u(cxc^{-1})u^{-1} = uxu^{-1}$$

We see that every element of $uC_G(x)$ produces the same element uxu^{-1} of cl(x). Thus, we can map the set $G/C_G(x)$ of left cosets to cl(x), by sending

$uC_G(x)$ to uxu^{-1}. This mapping is onto, since yxy^{-1} is hit by $yC_G(x)$. The map is also one to one, since if $uC_G(x)$ and $vC_G(x)$ map to the same element, then
$$uxu^{-1} = vxv^{-1} \quad \text{so that} \quad v^{-1}ux = xv^{-1}u$$
and then $v^{-1}u \in C_G(x)$, so that $u \in vC_G(x)$, which shows that $uC_G(x) = vC_G(x)$ (we do not necessarily have $u = v$!). □

The conjugacy classes in a group G that have exactly one element are the classes for elements of the center $Z(G)$. Let $L_1, \ldots, L_k$ be the distinct conjugacy classes which have at least two elements. Proposition 13.1 shows that the distinct classes are disjoint, and so

(13.1) $$|G| = |Z(G)| + \sum_{i=1}^{k} |L_i|$$

(If G is abelian, we let $k = 0$ and the sum on the right is vacuously 0.) Equation (13.1) is often called the *class equation* of the group G.

SYLOW'S THEOREM. *Let q be a power of a prime, and suppose that q divides $|G|$ where G is a finite group. Then G has at least one subgroup of order q.*

PROOF. Let G be a minimal counterexample, and find a prime p and $q = p^e$ where q divides $|G|$ but G has no subgroup of order q. Since G does have a subgroup of order 1, we see that $e \geq 1$ and $q \geq p$.

We consider equation (13.1), and we ask whether p divides each of the $|L_i|$. Suppose there is i such that p **does not** divide $|L_i|$. The set L_i is $\text{cl}(x)$ for some $x \in G$, and Proposition 13.2 shows that $|L_i| = |G/C_G(x)|$. Since, by definition, $|L_i| > 1$, we have that $|C_G(x)| < |G|$. Also, we have that p^e divides $|G| = |G/C_G(x)| \cdot |C_G(x)|$, and p does not divide $|G/C_G(x)| = |L_i|$. Therefore, $q = p^e$ divides $|C_G(x)|$. By the minimality of G, the group $C_G(x)$ has a subgroup of order q, but this is a subgroup of G of order q, a contradiction.

We must have that p divides every $|L_i|$. Since p divides $|G|$, equation (13.1) shows that p divides $|Z(G)|$. The group $Z(G)$ is abelian, and so Cauchy's Theorem of Abelian Groups gives it a subgroup H of order p. Since H is a subgroup of $Z(G)$, we have that $H \triangleleft G$. We have

$$|G| = |G/H| \cdot |H| = |G/H| \cdot p$$

Thus, $|G/H|$ is divisible by p^{e-1}. The group G/H has order less than that of G, and the minimality of G produces a subgroup J of G/H such that $|J| = p^{e-1}$. By the Correspondence Theorem, we have $J = K/H$ where H is a subgroup of K and K is a subgroup of G. Then

$$|K| = |K/H| \cdot |H| = |J| \cdot |H| = p^{e-1} \cdot p = q$$

This is a contradiction. □

When the prime power q of the hypothesis of Sylow's Theorem is just a prime, this result is called *Cauchy's Theorem*; it generalizes Cauchy's Theorem for Abelian Groups. When q is as large as possible a prime power that divides $|G|$, the subgroup produced is called a *Sylow subgroup*.

Sylow's Theorem is the beginning of the advanced study of the structure of finite groups. We will do applications of it in class. Here is one such. Notice how many previous results are brought together in the argument; we have chosen this result because it exhibits this sort of structure making it typical of group theoretic proofs.

PROPOSITION 13.3. *Let G be a finite group of order pq where $p < q$ are primes. Then G has a normal subgroup of order q. If, additionally, p does not divide $q - 1$, then G is cyclic.*

PROOF. Sylow's Theorem shows that G has a subgroup Q of order q. Then $|G/Q| = p$, and p is the smallest prime divisor of $|G|$. By a problem in Chapter 12, we see that $Q \triangleleft G$.

Sylow's Theorem gives G a subgroup of order p. Since p is prime, this subgroup is generated by an element y of order p. Proposition 12.8 showed that the mapping $f : Q \to G$ defined by $f(z) = yzy^{-1}$ is an isomorphism. Since Q is a normal subgroup of G, Proposition 11.1 shows that f maps Q into itself, and therefore f maps Q *onto* itself. We see that f is an automorphism of Q. Since Q is cyclic, Proposition 12.7 describes its automorphism group as isomorphic to U_q. We know that U_q has order $q-1$, and so Corollary 10.2 shows that the order of f divides $q-1$.

We claim that the order of f divides p. Indeed, compute that $f^k(z) = y^k z y^{-k}$, for each positive integer k. Since $y^p = 1$, we have that $f^p(z) = z$ for all $z \in Q$. We see that f^p is the identity automorphism, and so the order of f divides p.

If p does not divide $q-1$, then since the order of f divides both $q-1$ and p, it must be 1. In other words, f is the identity function and $x = f(x) = yxy^{-1}$. This shows that x and y commute and it follows that xy has order pq, so that G is cyclic. □

You will use the class equation to prove the following surprising result.

PROPOSITION 13.4. *Let G be a group of order q, where q is a power of an integer prime, and assume that $|G| > 1$. Then $|Z(G)| > 1$.*

We give two applications of Proposition 13.4. First, here is a simple proof of Proposition 12.4: let G be a group of order p^2, where p is prime. Could there be $x \in G \setminus Z(G)$? If so, then $Z(G) \subseteq C_G(x)$, and $x \in C_G(x)$, so that $|C_G(x)| > |Z(G)|$. Proposition 13.4 shows that the latter number is at least p, and so $|C_G(x)|$ is at least p^2. In other words, $x \in Z(G)$, a contradiction. We have just proved that $G = Z(G)$, and so G is abelian.

Here is a second application of Proposition 13.4: groups of prime power order have lots of normal subgroups.

PROPOSITION 13.5. *Let G be a group of order q, where q is a power of a prime. If r is a positive integer dividing q, then G has a normal subgroup of order r.*

PROOF. Proof by contradiction: let G be a minimal counterexample. Write $|G| = p^k$ where p is an integer prime and k is a non-negative integer. If $k = 0$, then $|G| = 1$, so that $r = 1$, and $G \triangleleft G$ shows that G is not a counterexample. Thus, $k \geq 1$, so that $|G| > 1$. Choose $r = p^j$, where $0 \leq j \leq k$, and such that G has no normal subgroup of order r. Since G has normal subgroups of order 1 and p^k, we see that $1 < j < k$.

Proposition 13.4 shows that $|Z(G)| > 1$. Cauchy's Theorem for Abelian Groups finds $x \in Z(G)$ of order p, and we see that $\langle x \rangle \triangleleft G$ with $|\langle x \rangle| = p$. The group $G/\langle x \rangle$ has order $p^{k-1} < |G|$, and p^{j-1} divides the order of $G/\langle x \rangle$. By the minimality of G, the group $G/\langle x \rangle$ has a normal subgroup of order p^{j-1}. The Correspondence Theorem shows that this subgroup has the form $H/\langle x \rangle$ where H is a subgroup of G. Since $H/\langle x \rangle \triangleleft G/\langle x \rangle$, statement (c) of the Correspondence Theorem shows that $H \triangleleft G$. Lagrange's Theorem shows that

$$|H| = |H/\langle x \rangle| \cdot |\langle x \rangle| = p^{j-1} \cdot p = p^j$$

and now G has a normal subgroup of order $p^j = r$. This contradiction proves the result. □

Sylow's original research paper is worth reading. It is in the journal Mathematische Annalen for the year 1872, starting on p.588 of that issue. (The article is in English.)

We have begun to give you the flavor of group theory. For the purposes of math 300, we must take our leave of groups to pursue other important topics. However, we hope that you are starting to believe that abstract algebra has its own structure and beauty.

Problems

158. Write down the class equation for the groups D_8 and $\mathbb{S}_3$.

159. Prove Proposition 13.4. (Hint: class equation mod the prime!)

160. Can there be a finite group G with $G \backslash Z(G)$ consisting of one conjugacy class? (Prove not, or find an example.)

161. Let G be a group of order q, where q is a power of an integer prime. Let H be a subgroup of G with $H \neq G$. Show that $H \subset N(H)$. (Hint: let G be a minimal counterexample. Show that you can assume that $Z(G) \subseteq H$. Factor groups!)

162. Let G be a group with $|G| = 4 \cdot p^2$ where p is prime. Show that G is not simple. (Hint: Sylow and Proposition 12.2, but don't forget the case $p = 2$.)

163. Show that a group of order 45 cannot be simple.

164. Follow the outline given to show that a group G of order 63 cannot be simple.

(a) Let $x \in G$ have order 7. (Why does x exist?)

(b) Use Proposition 12.2 on the subgroup $\langle x \rangle$ to get a homomorphishm f. What is the cycle structure of $f(x)$?

(c) Show that there is $y \notin \langle x \rangle$ such that $x \cdot y \cdot \langle x \rangle = y \cdot \langle x \rangle$. Show that $\langle x \rangle \subset N(\langle x \rangle)$. Show that either $\langle x \rangle \triangleleft G$ or that $|N(\langle x \rangle)| = 21$.

(d) If $|N(\langle x \rangle)| = 21$, use Corollary 12.3.

165. Follow the outline given to show that a group of order $p \cdot q \cdot r$, where p, q, r are distinct primes, cannot be simple.

(a) Let G be a simple group of order $p \cdot q \cdot r$, where $p < q < r$ are primes. Let $x \in G$ have order p. (Why does x exist?)

(b) Then G cannot have a subgroup of order $p \cdot q$, nor of order $p \cdot r$. (Hint: Corollary 12.3.)

(c) Then $C_G(x) = \langle x \rangle$. Also $\langle x \rangle$ is the normalizer of $\langle x \rangle$ in G.

(d) There are at least $(p-1) \cdot q \cdot r$ elements of order p in G. (Hint: what are the conjugacy classes of the elements of $\langle x \rangle$?)

(e) Let $y \in G$ have order q. (Why does y exist?) Then $|\text{cl}(y)| < q \cdot r$. (Why?)

(f) q divides $|C_G(y)|$ and $|C_G(y)| > p$. (Why?)

(g) Case 1: $|C_G(y)| = q \cdot r$, in which case there is an element of order $q \cdot r$ and a unique subgroup of order $q \cdot r$, which must therefore be normal. Contradiction.

(h) Case 2: $|C_G(y)| \neq q \cdot r$. Then the size of $|C_G(y)|$ contradicts step (b).

CHAPTER 14

Commutative Rings

When we mentioned that each of the sets $\mathbb{Z}$ and $\mathbb{Z}_n$ is a group under addition, we complained about leaving out multiplication. The two operations on each set have similar properties and are connected by the distributive law. There are important differences, however. In the integers, if $ab = 0$, then $a = 0$ or $b = 0$, whereas in $\mathbb{Z}_4$, for example, $2 \cdot 2 \equiv 0$. Also, the integers has an ordering that is crucial to its properties, whereas $\mathbb{Z}_n$ has no such useful ordering. The following definition focusses on arithmetic and does not concern itself with order; it is general enough to include both the integers and the modular integers, as well as many other sets.

DEFINITION 14.1. *The set R is a* commutative ring *if it has two commutative operations, denoted $+$ and $\cdot$, such that*
(a) $(a + b) + c = a + (b + c)$ *for all $a, b, c \in R$;*
(b) *there is $0 \in R$ such that $a + 0 = a$ for all $a \in R$;*
(c) *for each $a \in R$, there is $b \in R$ such that $a + b = 0$;*
(d) $(a \cdot b) \cdot c = a \cdot (b \cdot c)$ *for all $a, b, c \in R$;*
(e) *there is $1 \in R$ such that $a \cdot 1 = 1 \cdot a = a$ for all $a \in R$;*
(f) $a \cdot (b + c) = (a \cdot b) + (a \cdot c)$ *for all $a, b, c \in R$.*

Cancellation is so important to solving equations that we want to identify the situation in which it holds. We say that a commutative ring R is a *domain* if it has the cancellation property: if $a, b \in R$ and $a \cdot b = 0$, then $a = 0$ or $b = 0$.

Many sets that you have seen in other courses are commutative rings – the rationals, the reals, the complex numbers. If we remove the requirement that multiplication be commutative, then the $n \times n$ matrices over the reals form a (non-commutative) ring. We will stick to commutative rings, but we want you to see that you already have worked with rings in a variety of contexts. For the immediate purpose of this course, we want to introduce some interesting rings, and we want to display the analogy with groups, homomorphisms, and normal subgroups. Ring theory is a vast subject and it is covered in much more detail in a course in advanced algebra.

Definition 14.1abc shows that a commutative ring is a group under its addition operation. Thus, the additive identity element 0 is unique, and, for each $a \in R$, the additive inverse mentioned in 14.1c is unique. We can write $0_R = 0$ when we are worried about confusing the additive identity of R with the integer 0. As you expect, we denote by $-a$ the additive inverse of a.

By the way, in $\mathbb{Z}_1$, we have $0 = 1$, and the ring with only one element is the only ring in which $0 = 1$. Some people require $0 \neq 1$ for a ring, so that, as far as they are concerned, $\mathbb{Z}_1$ is not a ring. We will not need to make this distinction.

Here are some simple facts that follow from Definition 14.1. You already know that these facts are true in the examples of rings that you have seen.

PROPOSITION 14.2. *Let R be a commutative ring. Then the multiplicative identity element (from 14.1e) is unique. Furthermore, $a \cdot 0 = 0 \cdot a = 0$ and $(-1) \cdot a = -a = a \cdot (-1)$ for all $a \in R$.*

PROOF. If $e, 1 \in R$ are multiplicative identity elements, then Definition 14.1e shows that
$$1 = e \cdot 1 = e$$
Thus, 1 is the unique multiplicative identity.

Use Definition 14.1f to calculate

$$a \cdot 0 = a \cdot (0 + 0) = a \cdot 0 + a \cdot 0$$

Adding $-(a \cdot 0)$ to both sides gives $0 = a \cdot 0$. Similarly, $0 \cdot a = 0$.

Notice that -1 is, by definition, the additive inverse of 1, so that

$$(-1) + 1 = 0 = 1 + (-1)$$

Then use Definition 14.1f to calculate

$$(-1) \cdot a + a = (-1) \cdot a + 1a = ((-1) + 1) \cdot a = (0) \cdot a = 0$$

By the uniqueness of the additive inverse, we see that $(-1)a = -a$. Similarly $a \cdot (-1) = -a$. □

We write $1_R = 1$ to indicate clearly the multiplicative identity element of the ring R.

Here is an interesting ring to play with. Define $R_2 = \mathbb{Z} \times \mathbb{Z}$ with these operations.

$$(a_1, a_2) + (b_1, b_2) = (a_1 + b_1, a_2 + b_2)$$
$$(a_1, a_2) \cdot (b_1, b_2) = (a_1 \cdot b_1 + 2 \cdot a_2 \cdot b_2 \,,\, a_1 \cdot b_2 + a_2 \cdot b_1)$$
$$\text{for all} \quad a_1, a_2, b_1, b_2 \in \mathbb{Z}$$

It is easy (and tedious) to verify that R_2 is a commutative ring under these operations. For instance

$$0_{R_2} = (0, 0) \quad \text{and} \quad 1_{R_2} = (1, 0)$$

Here are two interesting properties of this ring. First, notice that the elements $(a, 0)$ behave like the integers:

$$(a, 0) + (b, 0) = (a + b, 0) \quad \text{and} \quad (a, 0) \cdot (b, 0) = (a \cdot b, 0)$$

In other words, it looks like R_2 contains a copy of the integers, Second, notice that
$$(0,1) \cdot (0,1) = (2,0)$$
The element on the right corresponds to the integer 2. It looks as if we have a square root of 2 in R_2; in fact, R_2 can be used to *construct* the square root of 2 from the integers. Where are the decimal places of $\sqrt{2}$?, you might ask. Good question.

The elements of a ring that have multiplicative inverses are the *units* of the ring. The units of the integers are 1 and -1; the units of $\mathbb{Z}_n$ are the elements of U_n. The following merely rehearses the definition of a group. Its proof is left to you.

PROPOSITION 14.3. *The set of units of a commutative ring R forms an abelian group under the multiplication of the ring.*

For a unit u in R, we write u^{-1} for the multiplicative inverse of u. We still wish to avoid fraction notation, and so we will **not** write $1/u$ for u^{-1}.

Whether an element is a unit may depend on context. The integer 2 is not a unit in the integers, since $1/2$ is not an integer. But 2 *is* a unit in the rational numbers.

What are the units in R_2? Check that $(1,1)$ is a unit! There are many others in this ring. Can you show that there are infinitely many?

We need another definition. A *field* is a commutative ring in which every non-zero element has a multiplicative inverse. Another way to say this: a field is a commutative ring in which all non-zero elements are in the units group.

PROPOSITION 14.4. *A field is a domain.*

PROOF. Let a and b be an element of the field R, and suppose that $a \cdot b = 0$. If $a \neq 0$, then a has a multiplicative inverse. Then $b = a^{-1} \cdot a \cdot b = a^{-1} \cdot 0 = 0$. □

The integers is a domain that is not a field (since 2, for example, does not have a multiplicative inverse in the integers). Also, R_2 is a domain that is not a field. The situation in $\mathbb{Z}_n$ is simple and very interesting.

PROPOSITION 14.5. *Let n be an integer greater than 1. If n is prime, then $\mathbb{Z}_n$ is a field. If n is not prime, then $\mathbb{Z}_n$ is not a domain.*

PROOF. The first part of the proof is Proposition 8.2. We repeat the argument here to make that argument more basic. If n is prime, and if $a \in \mathbb{Z}_n$ is not zero, then n does not divide a, and so the GCD of n and a is 1. By the Division Theorem, we can write $1 = a \cdot b + c \cdot n$ for some integers b, c. Reading this in $\mathbb{Z}_n$, we have

$$1 \equiv a \cdot b + c \cdot n \equiv a \cdot b$$

This proves that a has a multiplicative inverse.

If n is not prime, then since $n > 1$, there are natural numbers a, b, both less than n, such that $n = a \cdot b$. We have $a \not\equiv 0$ and $b \not\equiv 0$, but

$$a \cdot b \equiv n \equiv 0$$

so that $\mathbb{Z}_n$ is not a domain. □

In Chapter 11 we defined the *normal subgroup* as a special subgroup whose cosets form a group. In a following chapter, we defined *group homomorphism* and showed that the *kernel* of a homomorphism is a normal subgroup, and each normal subgroup is the kernel of a (natural) homomorphism. We also mentioned that the flow of subject could be reversed, beginning with homomorphisms and discovering normal subgroups. We will undertake that order of presentation here.

The definition of *ring homomorphism* is analogous to the definition of group homomorphism: if R, S are commutative rings, then $f : R \to S$ is a ring

homomorphism if
$$f(a+b) = f(a) + f(b)$$
$$f(a \cdot b) = f(a) \cdot f(b) \quad \text{for all} \quad a, b \in R$$

Thus, f intertwines both operations, addition and multiplication, of the rings. Notice that if we limit to addition, then f is a *group homomorphism* of R to S. Thus, $\ker(f)$ is defined as the set of $a \in R$ such that $f(a) = 0_S$. Proposition 12.1 shows that $\ker(f)$ is a normal subgroup of R. Since R is an abelian group under addition, every subgroup is normal, so normality is not an insight. How does the kernel relate to multiplication?

PROPOSITION 14.6. *Let R, S be commutative rings, and let $f : R \to S$ be a ring homomorphism. If $x \in \ker(f)$ and $r \in R$, then $x \cdot r \in \ker(f)$.*

PROOF. Compute:
$$f(x \cdot r) = f(x) \cdot f(r) = 0 \cdot f(r) = 0$$

□

Proposition 14.6 suggests the following definition: a subset I of R is called an *ideal*[1] of R if I is a subgroup of R under addition, and if $x \in I$ and $r \in R$ implies that $x \cdot r \in I$. Ideals in rings are analogous to normal subgroups of groups. Normal subgroups give factor groups. Ideals give *factor rings*.

PROPOSITION 14.7. *Let I be an ideal of the commutative ring R. Then R/I is a group under addition of cosets. We can define coset multiplication by the formula*
$$(a + I) \cdot (b + I) = (a \cdot b) + I \quad \text{for all} \quad a, b \in R$$

With these operations, the set R/I of cosets is a commutative ring. Its additive identity element is the coset I; its multiplicative identity is the coset $1 + I$.

---

[1]The word *ideal* was suggested by Dedekind; see the note at the end of this chapter.

PROOF. Group theory tells us that R/I is a subgroup under coset addition. We need to show that multiplication can be defined by the formula given. Here is the key fact: for $a, b \in R$, let $x \in a+I$ and $y \in b+I$. Then $x \cdot y \in (a \cdot b)+I$. Indeed, write $x = a+r$ and $y = b+s$, where $r, s \in I$, and compute

$$x \cdot y = (a+r) \cdot (b+s) = a \cdot b + a \cdot s + r \cdot b + r \cdot s$$

Each of the products as, rb, rs has at least one element of I. By the definition of ideal, each of these products is in I. Since I is a subgroup under addition, the sum of these elements is in I. This proves the claim.

The claim shows that if we are given cosets P, Q, there is a unique coset containing the products $p \cdot q$, where $p \in P$ and $q \in Q$. We define $P \cdot Q$ to be this coset. The claim shows that coset multiplication looks like this:

$$(a+I) \cdot (b+I) = (a \cdot b) + I \quad \text{for all} \quad a, b \in R$$

and this is the formula we need.

The statements in Definition 14.1 follow easily. $\square$

Every normal subgroups of a group is the kernel of the canonical homomorphism onto the normal subgroup's factor group. So with rings. If I is an ideal of the ring R, we can define $f : R \to R/I$ by $f(x) = x + I$. Proposition 14.7 shows that f is a ring homomorphism, and it is easy to see that its kernel is I.

We will give examples of ideals and ring homomorphisms via the exercises.

The theory of rings and ideals arose in number theory and in the theory of functions during the 1800's. Harold Edwards' article, The Genesis of Ideal Theory, in *Archive for History of Exact Sciences*, vol 23, 1980, pp.321-378, contains a thorough discussion. Ideals, like normal subgroups, foster an extremely effective, abstract approach to a variety of problems. As we mentioned, there is a great deal of interesting material here for an advanced course.

Problems

166. Let R and S be commutative rings, and define addition and multiplication on $R \times S$ as follows:
$$(r_1, s_1) + (r_2, s_2) = (r_1 + r_2, s_1 + s_2)$$
$$(r_1, s_1) \cdot (r_2, s_2) = (r_1 \cdot r_2, s_1 \cdot s_2)$$
$$\text{for all} \quad r_1, r_2 \in R, \ s_1, s_2 \in S$$

Show that Definition 14.1(b),(d),(e),(f) hold. (Actually, all the parts of Definition 14.1 hold, so that $R \times S$ is a commutative ring; it is the *ring direct product* of R and S. Notice that even though $R_2 = \mathbb{Z} \times \mathbb{Z}$ as a set, R_2 is **not** a ring direct product.)

167. Let R and S be commutative rings, each with at least two elements. Show that the ring $R \times S$ cannot be a domain.

168. Consider the ring R_2 defined above. Show that (a, b) is a unit in R_2 if and only if $a^2 - 2 \cdot b^2 = \pm 1$. Find a, b so that $a^2 - 2 \cdot b^2 = -1$ and find a, b such that $a^2 - 2 \cdot b^2 = 1$.

169. Let p be an integer prime. Define $R_p = \mathbb{Z} \times \mathbb{Z}$ with operations:
$$(a_1, a_2) + (b_1, b_2) = (a_1 + b_1, a_2 + b_2)$$
$$(a_1, a_2) \cdot (b_1, b_2) = (a_1 \cdot b_1 + p \cdot a_2 \cdot b_2, a_1 \cdot b_2 + a_2 \cdot b_1)$$
$$\text{for all} \quad a_1, a_2, b_1, b_2 \in \mathbb{Z}$$

Assume that these operations satisfy 14.1a,c,d,f.

(a) Show that 14.1b,e hold. (Note: in R_p we have $(0, 1)^2 = (p, 0)$, so R_p has a *square root* of p. Also note that R_p is not a ring direct product.)

(b) Show that R_p is a domain. (Hint: show that $(a_1, a_2) \cdot (a_1, -a_2)$ corresponds to a non-negative integer.)

170. Let R be a finite domain. Let $x \in R$ with $x \neq 0$. Define $f : R \to R$ by $f(y) = x \cdot y$. Show that f is one to one. Conclude that f is onto. (Why?) Conclude that there is $y \in R$ such that $x \cdot y = 1$. You have just proved that a finite domain is a field.

171. For a positive integer n, define $C_n = \mathbb{Z}_n \times \mathbb{Z}_n$ with addition and multiplication as follows:

$$(a_1, a_2) + (b_1, b_2) = (a_1 + b_1, a_2 + b_2)$$
$$(a_1, a_2) \cdot (b_1, b_2) = (a_1 \cdot b_1 - a_2 \cdot b_2, \ a_1 \cdot b_2 + a_2 \cdot b_1)$$
$$\text{for all} \quad a_1, a_2, b_1, b_2 \in \mathbb{Z}$$

Assume that C_n is a commutative ring.

(a) Find the multiplicative identity element.
(b) Find a copy of $\mathbb{Z}_n$ in C_n, and show that there is an element whose square is -1.
(c) Show that C_3 is a field.
(d) Show that C_5 is not a field.

172. Let I and J be ideals of the commutative ring R. Show that $I \cap J$ is an ideal of R.

173. Let I be an ideal of the commutative ring R, and let J be an ideal of the commutative ring S. Show that $I \times J$ is an ideal of the direct product $R \times S$.

174. Let I and J be ideals of the commutative ring R, and suppose that $I \cap J = \{0\}$. Show that $x \cdot y = 0$ or all $x \in I$ and $y \in J$.

175. Let R be a commutative ring and let $t \in R$. Show that the set $t \cdot R$ of $t \cdot r$ for all $r \in R$, is an ideal of R.

176. Let p be an integer prime, and consider the ring R_p defined above. Define $f : R_p \to R_p$ by $f(a,b) = (a,-b)$. Show that f is ring homomorphism. What is $\ker(f)$?

177. Let I be an ideal of $\mathbb{Z}$. Show that $I = n \cdot \mathbb{Z}$ for some $n \in \mathbb{Z}$, by completing the following steps.

(a) If $I = \{0\}$, we are done.

(b) Let $I \neq \{0\}$. Show that $I \cap \mathbb{N}$ is not empty, and let $n \in I \cap \mathbb{N}$ be minimal.

(c) Let $m \in I$, and use the Division Theorem to show that $m = q \cdot n$ for some $q \in \mathbb{Z}$.

178. Let $m, n \in \mathbb{Z}$. Show that $m \cdot \mathbb{Z} \subseteq n \cdot \mathbb{Z}$ if and only if n divides m.

179. Let I be an ideal of the commutative ring R. Define $f : R \to R/I$ by $f(x) = x + I$. Show that f is a ring homomorphism and that $\ker(f) = I$.

180. An ideal I of the ring R is *maximal* if $I \neq R$ and whenever J is an ideal with $I \subseteq J \subseteq R$, we have either $J = I$ or $J = R$. Let $p \in \mathbb{Z}$ be prime. Show that $p \cdot \mathbb{Z}$ is a maximal ideal of $\mathbb{Z}$. (Hint: previous problems have shown what are the ideals of $\mathbb{Z}$ and when it is that one contains another.)

181. Let I, J be ideals of the commutative ring R. Show that $I + J$ is an ideal of R. Show that if $i \in I$ and $j \in J$ and $a, b \in R$, then $a \cdot i + b \cdot j \in I + J$.

182. Let Z be the subset of R_2 consisting of $(a, 0)$, where $a \in \mathbb{Z}$, so that Z looks like a copy of the integers.

(a) Let I be an ideal of R_2. Show that $I \cap Z$ is an ideal of Z.

(b) Let $J = (0,1) \cdot R_2$. Show that J consists of (a,b) such that p divides a.

183. Let R be a commutative ring with exactly two ideals: R and $\{0\}$. Show that R is a field. (Hint: for $x \in R \setminus \{0\}$, think about $x \cdot R$.)

CHAPTER 15

Residue Units

By this point in the course, we have covered a selection of the standard topics. We could end our study with any one of a number of investigations; we have chosen to pursue some number theory, both to cover a result relevant to combinatorics and computer science and to review some simple group theory.

We know that the set U_n of units in $\mathbb{Z}_n$ is the set of $x \in \mathbb{Z}_n$ such that the GCD of x and n is 1. Back in Chapter 9 we introduced Euler's totient function $\phi(n)$, we know that $\phi(n) = |U_n|$. We will make an initial study of the structure of U_n. To get off the ground, we need the following.

CHINESE REMAINDER THEOREM. *Let m and n be integers with GCD 1. Choose integers a and b. Then there is an integer x such that $x - a$ is divisible by m and $x - b$ is divisible by n.*

PROOF. Since m and n have GCD equal to 1, we can find integers c and d such that $cn + dm = 1$. We wish to multiply this equation by $b - a$. Let $p = (b-a)c$, let $q = (b-a)d$, and then

$$pn + qm = b - a \quad \text{so that} \quad qm + a = b - pn$$

Let $x = qm + a$, and it is obvious that $x - a$ is divisible by m and $x - b$ is divisible by n. □

Now we can move toward an interesting formula for the order of U_n (for the totient function).

PROPOSITION 15.1. *Let n and m be positive integers with GCD 1. Then $\phi(nm) = \phi(n) \cdot \phi(m)$.*

PROOF. We will construct a one to one, onto function

$$\Psi : U_{mn} \to U_n \times U_m$$

This map will, in fact, be a group isomorphism, although we will not need or prove that fact.

To construct Ψ, we need to consider the modulii m and n and mn all at the same time. For $x \in \mathbb{Z}$, we write x_n when x is considered as an element of $\mathbb{Z}_n$. Similarly for m and mn in place of n.

To give the definition of Ψ, we first let its domain be the integers. For $x \in \mathbb{Z}$, define

$$\Psi(x) = (\, x_m \,,\, x_n \,)$$

and this gives us $\Psi : \mathbb{Z} \to \mathbb{Z}_m \times \mathbb{Z}_n$.

We claim that if $x \equiv y \mod mn$, then $\Psi(x) = \Psi(y)$. Indeed, we have $x = y + cmn$ for some integer c. It is easy to see that $x \equiv y \mod m$ and $x \equiv y \mod n$, and this shows that $x_m = y_m$ and $x_n = y_n$. Thus, $\Psi(x) = \Psi(y)$, as claimed.

The previous paragraph shows that we can regard Ψ as having domain $\mathbb{Z}_{mn}$ (rather than $\mathbb{Z}$). Next we show that if $x \in U_{mn}$, then $\Psi(x) \in U_m \times U_n$. In other words, we need to prove that $x_m \in U_m$ and that $x_n \in U_n$, assuming that $x \in U_{mn}$. Well, if g is the GCD of x and m, then g divides mn as well. Since the GCD of x and mn is 1, this implies that g divides 1. This proves that the GCD of x and m is 1, and so $x_m \in U_m$. Similarly, $x_n \in U_n$, and we have that Ψ maps U_{mn} into $U_m \times U_n$.

Next we prove that Ψ maps U_{mn} *onto* $U_m \times U_n$. Let $a_m \in U_m$ and $b_n \in U_n$. By the Chinese Remainder Theorem, there is an integer x such that $x - a$ is divisible by m (so that $x_m \equiv a_m$) and $x - b$ is divisible by n (so that $x_n \equiv b_n$).

We see that
$$\Psi(x_{mn}) = (x_m, x_n) = (a_m, b_n)$$
We have left to prove that $x_{mn} \in U_{mn}$. If this is false, then the GCD of x and mn has a prime divisor p. Since p divides mn, it divides m or it divides n. Suppose, for example, that p divides m. We had that p divides x, as well, and so $x_m \notin U_m$. Recall that $x_m \equiv a_m$, and since $a_m \in U_m$, we see that $x_m \in U_m$, and this is a contradiction. Thus, $x_{mn} \in U_{mn}$.

It remains to show that Ψ is one to one. Assume that $\Psi(x_{mn}) = \Psi(y_{mn})$. Then the coordinates must be equivalent:
$$x_m \equiv y_m \quad \text{and} \quad x_n \equiv y_n$$
Thus, both m and n divide $x - y$. Since the GCD of n and m is 1, Proposition 8.5 shows that mn divides $x - y$. Thus, $x_{mn} \equiv y_{mn}$. This proves that Ψ is one to one.

The multiplication formula stated in the present proposition follows easily. We have
$$\phi(mn) = |U_{mn}| = |U_m \times U_n| = |U_m| \cdot |U_n| = \phi(m) \cdot \phi(n)$$
$\square$

We mentioned that Ψ is a group isomorphism, and so there is group theoretic information implicit in Proposition 15.1. Our immediate concern, however, is with a multiplicative formula for $\phi(n)$.

COROLLARY 15.2. *Let n be a positive integer, and suppose that we find distinct primes $p_1, \ldots, p_r$ such that $n = \prod_{i=1}^{r} p_i^{e_i}$. Then*
$$\phi(n) = \prod_{i=1}^{r} \phi(p_i^{e_i})$$
Furthermore, if p is a prime and $e \geq 1$, then $\phi(p^e) = (p-1) \cdot p^{e-1}$.

PROOF. We apply Proposition 15.1 repeatedly to obtain the product formula for $\phi(n)$. It remains to prove the other equation.

An exercise on p.95 shows that U_{p^e} has order $(p-1) \cdot p^{e-1}$. □

We turn to some number-theoretic facts.

FERMAT'S LITTLE THEOREM. *Let n be a positive integer, let x be an integer, and suppose that the GCD of x and n is 1. Then $x^{\phi(n)} - 1$ is divisible by n. If n is prime, and x is an integer, then $x^n - x$ is divisible by n.*

PROOF. The first statement is an application of Corollary 10.2 to the group U_n. For the second statement, if $x \in U_n$, then since $\phi(n) = n - 1$, the first statement says that $x^{n-1} - 1$ is divisible by n. Multiplying by x, we get that $x^n - x$ is divisible by n. In the case that $x \notin U_n$, we must have that n divides x, since n is prime. We see easily that $x^n - x$ is divisible by n. □

Fermat's Little Theorem has many consequences. In the problems, we outline a commonly used data encryption scheme that arises from it.

We explore the case where n is a prime power in more detail. It turns out that if n is a power of an *odd* prime, then U_n is cyclic, as we will prove. This fact was known to Fermat, and it has consequences in the theory of polynomial equations and in many other areas. Our proof is essentially that of Gauss in the *Disquisitiones Arithmeticae* referred to on p.25. We begin with a couple lemmas; the first is the remainder theorem for polynomials.

PROPOSITION 15.3. *Let $p \in \mathbb{N}$. Let $A_0, \ldots, A_n, \alpha \in \mathbb{Z}_p$. Then there are $B_0, \ldots, B_{n-1} \in \mathbb{Z}_p$ such that*

$$\sum_{j=0}^{n} A_j \cdot \beta^j = (\beta - \alpha) \cdot \sum_{j=0}^{n-1} B_j \cdot \beta^j + \sum_{j=0}^{n} A_j \cdot \alpha^j \quad \text{for all} \quad \beta \in \mathbb{Z}_p$$

PROOF. Compute

$$\sum_{j=0}^{n} A_j \cdot \beta^j - \sum_{j=0}^{n} A_j \cdot \alpha^j = \sum_{j=0}^{n} A_j \cdot (\beta^j - \alpha^j)$$

$$= \sum_{j=1}^{n} A_j \cdot \left[(\beta - \alpha) \cdot \sum_{i=0}^{j-1} \beta^i \cdot \alpha^{j-1-i}\right]$$

$$= (\beta - \alpha) \cdot \sum_{i=0}^{n-1} \left[\sum_{j=i+1}^{n} A_j \cdot \alpha^{j-1-i}\right] \cdot \beta^i$$

We define

$$B_i = \sum_{j=i+1}^{n} A_j \cdot \alpha^{j-1-i} \quad \text{for} \quad 0 \le i \le n-1$$

$\square$

PROPOSITION 15.4. *Let p be a positive integer prime, and let $\alpha \in \mathbb{Z}_p$ have multiplicative order n. Then $\mathbb{Z}_p$ has exactly n elements whose n-th power is 1.*

PROOF. If $n = 1$, then the result is trivial, for only 1 has order 1. Assume that $n \ge 2$. Let $\beta \in \mathbb{Z}_p$ with $\beta^n = 1$, and we will show that β is one of the n powers of α. Assume, to the contrary, that $\beta \ne \alpha^j$ for all j.

We have that $\beta^n - 1 = 0$ and $(\alpha^j)^n - 1 = 0$ for all j with $0 \le j < n$. Choose the positive integer m minimal such that there are $A_0, \ldots, A_m \in \mathbb{Z}_p$, with $A_0 \ne 0$ and such that

(15.1) $$\sum_{i=0}^{m} A_i \cdot \beta^i = 0 \quad \text{and} \quad \sum_{i=0}^{m} A_i \cdot (\alpha^j)^i = 0 \quad \text{for} \quad 0 \le j < m$$

We apply Proposition 15.3, with α^{m-1} in place of the α of that proposition. There are $B_0, \ldots, B_{m-1}$ such that

$$\sum_{i=0}^{m} A_i \cdot \gamma^i = (\gamma - \alpha^{m-1}) \cdot \sum_{j=0}^{m-1} B_j \cdot \gamma^j + \sum_{i=0}^{m} A_i \cdot (\alpha^{m-1})^i \quad \text{for all} \quad \gamma \in \mathbb{Z}_p$$

By (15.1), the term on the far right is 0, and we have

$$(15.2) \quad \sum_{i=0}^{m} A_i \cdot \gamma^i = (\gamma - \alpha^{m-1}) \cdot \sum_{j=0}^{m-1} B_j \cdot \gamma^j$$

Letting $\gamma = 0$, we have $A_0 = -\alpha^{m-1} \cdot B_0$, and this shows that $B_0 \neq 0$.

In (15.2), let $\gamma = \alpha^k$ for $0 \leq k < m-1$. The sum on the left is 0. Since $\alpha^k \neq \alpha^{m-1}$, we see that

$$\sum_{j=0}^{m-1} B_j \cdot (\alpha^k)^j = 0$$

Let $\gamma = \beta$, and since $\beta \neq \alpha^{m-1}$, it follows that

$$\sum_{j=0}^{m-1} B_j \cdot \beta^j = 0$$

This contradicts the minimality of m, unless $m-1$ is not a positive integer – in other words $m = 1$. Equation (15.2) now looks like this:

$$A_0 + A_1 \cdot \gamma = (\gamma - 1) \cdot B_0 \quad \text{for all} \quad \gamma \in \mathbb{Z}_p$$

We know that $B_0 \neq 0$ and that $A_0 + A_1 \cdot \beta = 0$; we conclude that $\beta - 1 = 0$. This is a contradiction, for $1 = \alpha^n$. □

THEOREM 15.5. *If p is a prime, then the group U_p is cyclic.*

PROOF. Proposition 15.4 shows that Theorem 10.3 applies to U_p. □

We can generalize this to the case of an odd prime power. We will need the following, rather technical result.

LEMMA 15.6. *Let p be an odd prime and let e be a non-negative integer. Then $(1+p)^{p^e} = 1 + ap^{e+1}$ where p does not divide a. If e is a non-negative integer, then $5^{2^e} = 1 + a2^{e+2}$ where a is odd.*

15. RESIDUE UNITS

PROOF. Let p be an odd prime. The conclusion we want needs support from an even more technical result. We claim, for $0 \leq j$, that
$$(1+p)^j = 1 + jp + kp^2 + ip^3$$
where k, i are integers and $2k = j(j-1)$. (Many of you will recognize the Binomial Theorem here.) Indeed, this is true for $j = 0$ (using $k = i = 0$). For an inductive proof, assume that $(1+p)^j$ has the required form for $j \geq 0$, and consider

$$\begin{aligned}(1+p)^{j+1} = (1+p)^j(1+p) &= (1 + jp + kp^2 + ip^3)(1+p) \\ &= 1 + jp + kp^2 + ip^3 + p + jp^2 + kp^3 + ip^4 \\ &= 1 + (j+1)p + (k+j)p^2 + (i+k+ip)p^3\end{aligned}$$

Also,
$$2(k+j) = 2k + 2j = j(j-1) + 2j = (j+1)j$$
as needed.

Applying this result with $j = p$, we get
$$(1+p)^p = 1 + p^2 + kp^2 + ip^3$$
where $2k = p(p-1)$. Since p is odd, we see that p divides k, and so we can write $k = pc$. We have
$$(1+p)^p = 1 + p^2 + cp^3 + ip^3 = 1 + p^2(1 + cp + ip)$$
In other words, $(1+p)^p = 1 + p^2 a$ where p does not divide a.

The last result can be used to handle $(1+p)^{p^e}$. We prove the conclusion of the lemma by induction on e, the case $e = 1$ being what was just proved. Let $e \geq 2$, put $q = (1+p)^{p^{e-1}}$, so that by induction we have $q = 1 + bp^e$ where p does not divide b. Then $(1+p)^{p^e} = q^p$, so that
$$(1+p)^{p^e} - 1 = q^p - 1 = (q-1) \cdot \sum_{j=0}^{p-1} q^j$$

Using the identity for q, this is equal to
$$bp^e \cdot \sum_{j=0}^{p-1} q^j$$

We claim that the sum S on the right is divisible by p but not by p^2. If this is so, then that sum can be written pc where p does not divide c, and then we will have
$$(1+p)^{p^e} = 1 + bcp^{e+1}$$
as needed.

To consider the sum S, notice that it has p terms. Thus,
$$S - p = \sum_{j=0}^{p-1}(q^j - 1) = \sum_{j=1}^{p-1}(q-1)\sum_{i=0}^{j-1} q^i$$

Recall that $q - 1 = bp^e$ and that $e \geq 2$. Therefore, each term in $S - p$ is divisible by p^2, and we can write $S - p = dp^2$ for some integer d. Then $S = p + dp^2 = p(1 + dp^2)$, and this shows that S is divisible by p but not by p^2, as needed to complete the proof of the first statement in the present proposition.

Now we need to consider 5^{2^e}. For $e = 0$, this is $5 = 1 + 1 \cdot 4$. To use induction, let $e \geq 1$ and compute
$$5^{2^e} = \left(5^{2^{e-1}}\right)^2 = (1 + a2^{e+1})^2$$
where a is odd. Continuing,
$$(1 + a2^{e+1})^2 = 1 + a2^{e+2} + a^2 2^{2e+2} = 1 + (a + a^2 2^e)2^{e+2}$$

Since a is odd and $e \geq 1$, we see that $5^{2^e} = 1 + b2^{e+2}$, where b is odd. $\square$

THEOREM 15.7. *If q is a power of an odd prime, then U_q is cyclic. If $q = 2^e$ with $e \geq 2$, then U_q is isomorphic to the direct product of its cyclic subgroups $\langle -1 \rangle$ and $\langle 5 \rangle$, the former having order 2 and the latter 2^{e-2}.*

PROOF. Let $q = p^e$ where p is an odd prime and e is a positive integer. We first show that $1+p$ has order p^{e-1} in U_q. By Lemma 15.6, $(1+p)^{p^{e-1}} \equiv 1$. Thus, the order of $1+p$ in U_q divides p^{e-1}, and so this order is a power of p. Lemma 15.6 shows that p^{e-1} is the smallest power of p which raises $1+p$ to a number equivalent to 1 in $\mathbb{Z}_q$. Thus, $1+p$ has order p^{e-1} in U_q.

Next we show that there is an element in U_q of order $p-1$. The group U_p is cyclic (by Theorem 15.5), and so there is an integer a such that a has order $p-1$ in U_p. We claim that a has order divisible by $p-1$ in U_q. Suppose that m is the order of $a \in U_q$. Then q divides $a^m - 1$, and so p divides $a^m - 1$, and we see that $a^m \equiv 1$ in U_p. By Corollary 7.3, the number m must be divisible by $p-1$, as claimed. Since $p-1$ divides the order of a in U_q, this group has an element of order $p-1$, by Proposition 9.3.

You have done a problem where you showed that if x and y are commuting elements of a group having orders with GCD equal to 1, then xy has order $o(x)o(y)$. We see that U_q has elements of orders $p-1$ and p^{e-1}, therefore it has an element of order $(p-1)p^{e-1}$. The formula of Corollary 15.2 on the order of U_q now allows the conclusion that U_q is cyclic.

We are left with $q = 2^e$ where $e \geq 2$. That the element 5 has order 2^{e-2} follows from Lemma 15.6. The group $\langle -1 \rangle$ has order 2. We will prove that it intersects the group $\langle 5 \rangle$ in only the identity element. The direct product of these two cyclic groups has order 2^{e-1}, the order of U_q. This will complete the proof (recall Proposition 12.10).

We need to consider an element in the intersection of $\langle -1 \rangle$ and $\langle 5 \rangle$. Since $\langle -1 \rangle$ has only two elements, we ask whether -1 is an element of $\langle 5 \rangle$. Suppose that there is a positive integer m such that $-1 \equiv 5^m$ Then 2^e divides $5^m + 1$, and since $e \geq 2$, this implies that 4 divides $5^m + 1$. Taking 4 as modulus, notice that $5 \equiv 1$, and so $5^m \equiv 1$. Thus, 4 does not divide $5^m + 1$. This contradiction completes the proof. □

At the end of Chapter 8 we mentioned a couple of books on number theory. The facts considered here are discussed in those books in a more general context.

Problems

184. For $n = 13, 17, 19, 23$, find x such that $\langle x \rangle = U_n$. (Of course, the x will be different for each of the four n's.)

185. Let p be an integer prime with $p \geq 3$, and suppose that -1 has a square root mod p. We will show that $p \equiv 1 \mod 4$. Let $x \in \mathbb{Z}_p$ with $x^2 \equiv -1 \mod p$, and show that x has order 4 in U_p. Conclude that $p \equiv 1 \mod 4$.

186. Let p be an integer prime with $p \equiv 1 \mod 4$. We will show that -1 has a square root mod p.

(a) Show that $(p-1)! \equiv -1 \mod p$ (Hint: pair each factor with its multiplicative inverse mod p.)

(b) Find $x \in \mathbb{Z}_p$ such that $x^2 \equiv (p-1)! \mod p$. (Hint: write $(p-1)! = 1 \cdot (p-1) \cdot 2 \cdot (p-2) \cdots$. You get a square with a bunch of minus signs.)

187. Let p, q be distinct primes and let $n = p \cdot q$. Let d, e be inverses of each other mod $\phi(n)$. Show that $x^{d \cdot e} \equiv x \mod n$. (Hint: consider separately the case that p divides x, that q divides x and that neither divides x.) This problem can be made the basis of a coding scheme. The numbers n and d are public and e is known only to me. I may or may not know $\phi(n)$, but you do not know it. If you send me message x coded as x^d, I can decode it by raising it to the e-th power. Since only I know e, only I can decode. If I send you message x coded as x^e, you raise it to the d-th power to decode – that would ensure that the message came from me. This encryption scheme is called RSA, after Rivest, Shamir, and Adelman, who discovered it; it is described in more detail in many other resources.

CHAPTER 16

The Rational Numbers.

We have worked all term without fractions, concentrating attention on integer number theory. A typical advanced calculus course starts with the rational and real numbers, and so it is of obvious interest to construct the rationals at this point. We should mention, however, that as in Chapter 14, the real goal of this work is to further your contact with formal definitions. The first few results are stated without proof; we are leaving it to you to supply the details!

A rational number is an integer fraction m/n where $n \neq 0$, but we need to define *fraction*. This is not trivial, since fractions can be equal without having the same parts: $2/3 = 10/15$. The key to constructing the rationals is to formalize this ambiguity of form. We will see that this is analogous to wanting to treat 2 and 7, say, as equal elements of $\mathbb{Z}_5$.

We begin by defining a set of pairs that will eventually be turned into fractions. Let

$$F = \{(x, y) \mid x, y \in \mathbb{Z}, y \neq 0\}$$

The elements of F are ordered pairs (elements of $\mathbb{Z} \times \mathbb{Z}$). Thus, if $(a, b), (c, d) \in F$, then $(a, b) = (c, d)$ if and only if $a = c$ and $b = d$.

We define addition and multiplication on F; these operations will eventually give addition and multiplication of fractions, and the formulas we are about to give are motivated by that fact. For $(a, b), (c, d) \in F$, define

$$(a, b) + (c, d) = (ad + bc, bd) \quad \text{and} \quad (a, b) \cdot (c, d) = (ac, bd)$$

(Notice that $(a/b)+(c/d) = (ad+bc)/(bd)$. We're not quite ready for fractions, but we're close.)

In the definitions, the integers b, d cannot be 0, and so $bd \neq 0$, too, and this shows that the operations just defined are on F. These operations **do not** make F into a ring. However, many of the ring axioms do hold; here is a list.

PROPOSITION 16.1. *The addition and multiplication operations defined on F are commutative and they satisfy the following statements in Definition 14.1: (a), (b), (d), (e), (f). The zero element is $(0, 1)$ and the multiplicative identity is $(1, 1)$.*

The fractions a/b and c/d are equal if $ad = bc$, an integer equation. This equation is the key to turning the set F into a set of fractions. For each $(a, b), (c, d) \in F$, say $(a, b) \equiv (c, d)$ when $ad = bc$. We claim that this is an equivalence relation. Indeed, notice that $(a, b) \equiv (a, b)$, since $ab = ba$. Next, if $(a, b) \equiv (c, d)$, so that $ad = bc$, then also $(c, d) \equiv (a, b)$. We leave it to you to verify transitivity.

We construct the rational numbers from F the way we constructed $\mathbb{Z}_n$ from $\mathbb{Z}$. Define $\mathbb{Q}$ to be F, but with $\equiv$ replacing equality. Thus, $(2, 3)$ and $(-4, -6)$ are the same element of $\mathbb{Q}$. The set $\mathbb{Q}$ is the set of *rational numbers*. Now we can define fractions: for $a, b \in \mathbb{Z}$ with $b \neq 0$, we write $a/b = (a, b)$, and we define $a/b = c/d$ when $(a, b) \equiv (c, d)$. Our set $\mathbb{Q}$ of rational numbers can be taken to be the set of fractions.

The key property of fractions involves canceling. If $a, b, c \in \mathbb{Z}$ and $b \neq 0 \neq c$, then notice that

$$(a \cdot b, c \cdot b) \equiv (a, c) \quad \text{since} \quad a \cdot b \cdot c = c \cdot b \cdot a$$

In fractions, this is

$$\frac{a \cdot b}{c \cdot b} = \frac{a}{c}$$

We will use this repeatedly.

We need operations on $\mathbb{Q}$: we show that the addition and multiplication on F define an addition and multiplication on $\mathbb{Q}$. The argument is similar to that for Proposition 3.1. Note that each $x \in \mathbb{Q}$ is an element of F, so we already have an addition and multiplication in F.

PROPOSITION 16.2. *If $x, y, z \in \mathbb{Q}$ and if $x = y$ in $\mathbb{Q}$, then $x + z = y + z$ and $x \cdot z = y \cdot z$.*

PROOF. Let $x = a/b$ and $y = c/d$, so that $a \cdot d = b \cdot c$. Let $z = e/f$. Then
$$x + z = \frac{a \cdot f + b \cdot e}{b \cdot f} \quad \text{and} \quad y + z = \frac{c \cdot f + d \cdot e}{d \cdot f}$$
To show that $x + z = y + z$ we need equivalence of the relevant pairs in F:
$$(a \cdot f + b \cdot e) \cdot d \cdot f = (a \cdot d \cdot f + b \cdot e \cdot d) \cdot f$$
$$= (b \cdot c \cdot f + b \cdot e \cdot d) \cdot f \qquad \text{since } ad = bc$$
$$= b \cdot f \cdot (c \cdot f + d \cdot e)$$
as needed to show that $x + z = y + z$.

The argument for $x \cdot z = y \cdot z$ is similar. $\square$

Now we have addition and multiplication on $\mathbb{Q}$, and the properties on F transfer easily to $\mathbb{Q}$. Proposition 16.1 shows that Definition 14.1(a), (b), (d) hold. The proof is left to you. For instance, $1/1$ is the multiplicative identity element.

We use our newly minted fraction notation to prove the following.

PROPOSITION 16.3. *Under the addition and multiplication just defined, $\mathbb{Q}$ is a field.*

PROOF. We have left to verify Definition 14.1 (c), (f), and we need to show that each non-zero elements has a multiplicative inverse. From (b), note that $0/1$ is the additive identity on $\mathbb{Q}$, since $(0, 1)$ is the additive identity on F.

For (c), let $a/b \in \mathbb{Q}$, and compute that
$$(0, b \cdot b) \equiv (0, 1) \quad \text{since} \quad 0 \cdot 1 = b \cdot b \cdot 0$$
Thus $0/(b \cdot b) = 0/1$. Thus,
$$\frac{a}{b} + \frac{-a}{b} = \frac{a \cdot b - b \cdot a}{b \cdot b} = \frac{0}{b \cdot b} = \frac{0}{1}$$
This shows that a/b has an additive inverse.

For (f), let $a/b, c/d, e/f \in \mathbb{Q}$. compute that
$$\frac{a}{b} \cdot \left(\frac{c}{d} + \frac{e}{f}\right) = \frac{a}{b} \cdot \frac{c \cdot f + d \cdot e}{d \cdot f} = \frac{a \cdot c \cdot f + a \cdot d \cdot e}{b \cdot d \cdot f}$$
The right side of the equation in (f) is this.
$$\frac{a \cdot c}{b \cdot d} + \frac{a \cdot e}{b \cdot f} = \frac{a \cdot c \cdot b \cdot f + b \cdot d \cdot a \cdot e}{b \cdot d \cdot b \cdot f} = \frac{a \cdot c \cdot f + d \cdot a \cdot e}{d \cdot b \cdot f}$$

As for multiplicative inverses, if $a/b \neq 0/1$, then $a \neq 0$, and so $b/a \in \mathbb{Q}$. That's obviously the inverse. □

We also get an ordering on $\mathbb{Q}$. We define a/b to be *positive* when $a \cdot b > 0$. This is clever, since it tells us that numbers like $(-3)/(-2)$ are positive. You can prove that the sum and product of positive rational numbers is positive. For rational numbers $a/b, c/d$, we define $a/b < c/d$ exactly when $c/d - a/b$ is positive. This ordering on $\mathbb{Q}$ has properties similar to those for the integers, introduced in Chapter 2, as you can show.

Everything looks good, until we recall a very obvious feature of ordinary fractions. For each *integer* n, we expect to have
$$n = \frac{n}{1}$$
In other words, each integer should also be a rational number. Unfortunately, there is no hope of this in our construction, since the integer n is definitely not $n/1$, the latter object being several equivalent pairs of integers. On the

other hand, if we have any confidence in our construction (which was pretty natural), we might think that the set,

$$Z = \{n/1 \mid n \in \mathbb{Z}\}$$

which *should* be the integers, will have the same properties as the integers. In fact, that is the case. To be specific, the set Z (a subset of $\mathbb{Q}$) along with the addition, multiplication, and ordering defined for $\mathbb{Q}$, satisfies the integer axioms we gave back in Chapter 2, as you can check.

Now we have a problem. The original integers from Chapter 2 are not contained in the rationals, but the rationals do contain a set that looks exactly like the integers. When we do fraction arithmetic, which "integers" should we use? It is easy to see that we do not want to consider both sets of integers at once. Otherwise, fractions such as 6/2 might carry two separate meanings: $6/2 = 3$ expressing that $6 = 2 \cdot 3$ in $\mathbb{Z}$, and $6/3 = 3/1$ expressing an identity in the rationals. Most mathematicians get out of this dilemma by saying that they *identify* $\mathbb{Z}$ and Z. Formally, this means that we agree not to distinguish the "old integer" n from its rational version $n/1$, and thereby we widen the definition of equal on the integers.

Similarly, when you construct the real numbers from the rationals (often this is the way Math 310 begins), in order to get the rationals included in the set of real numbers, you have to pass to a subset of the reals, a subset that has the same properties as do the rationals, and the rationals are *identified* with this subset.

The downside of this approach is its uncertainty about what the integers really are. If numbers, for example, are *real* in the sense that they actually occur in the world, then "identifying" them in various contexts might cut us off from that reality. All this sounds philosophical because it is. Indeed, there is a lot of food for thought here.

We will adopt the usual approach that *identifies* the sets $\mathbb{Z}$ and Z. For all $n \in \mathbb{Z}$, we have $n = n/1$. For example, $0/1 = 0$ and $1/1 = 1$. Now also $\mathbb{Z} \subset \mathbb{Q}$.

There is a very important property of $\mathbb{Q}$ that moves us in the direction of *analysis* (the mathematics of limits) – namely that there is no smallest positive rational number. This idea is sometimes called the *Archimedean Property*. It will convenient for us to use powers of 2 to get small rational numbers in this proposition. We leave the proof to you.

PROPOSITION 16.4. *There is no smallest positive rational number. In fact, given $0 < r \in \mathbb{Q}$, there is a positive integer k such that $\frac{1}{2^k} < r$.*

The utility of having the rationals is that we can divide by integers. This is useful in solving equations and in expressing accurate measurements. Let us show, however, that the rationals are inadequate for both these purposes. Our first result is a generalization of Proposition 8.9 on the irrationality of the square roots of primes.

PROPOSITION 16.5. *Let n be a positive integer. Let m be an integer, and suppose that $x^n = m$ has no integer solution. Then $x^n = m$ has no rational solution.*

PROOF. Suppose, to the contrary, that $r^n = m$ for some rational number r, but for no integer r. Then $r \neq 0$, and if we write $r = a/b$ with $a, b \in \mathbb{Z}$, then since $a/b = (-a)/(-b)$, we can assume that $a > 0$.

Define
$$S = \{a \in \mathbb{N} \mid \exists b \in \mathbb{Z} \text{ with } (a/b)^n = m\}$$
The previous paragraph shows that S is not empty, and so, by well-ordering, it has a minimal element c. By definition of $c \in S$, there is $d \in \mathbb{Z}$ such that $(c/d)^n = m$, and this is

(16.1) $$c^n = md^n$$

We claim that d cannot be 1 or -1. If it were, then c/d would be an integer, contrary to the assumption on m. Thus, d has a prime divisor p. Equation (16.1) shows that p divides c^n, and so p divides c. Write $c = pc'$ and $d = pd'$, and note that $c' > 0$ and that $c' < c$. Substituting into (16.1) we get
$$p^n c'^n = m p^n d'^n$$
Cancelling the p^n we obtain
$$c'^n = m d'^n \quad \text{or} \quad \left(\frac{c'}{d'}\right)^n = m$$
Thus, $c' \in S$, yet $c' < c$ violates the minimality of c. This contradiction completes the proof. □

Proposition 16.5 shows that the rationals are no better than the integers for solving some kinds of equations. Let us hint at a way out of this problem by walking right up to the idea of the limit. We want to show that although a rational number might not have a rational n-th root, there are rational numbers arbitrarily close to the n-th root.

PROPOSITION 16.6. *Let r be a positive rational number, and let n be a positive integer. Let ϵ be a positive rational number. Then there is a rational number s such that s^n is within ϵ of r.*

PROOF. Let $s, t \in \mathbb{Q}$, and you can verify the identity
$$t^n - s^n = (t-s)\left(t^{n-1} + t^{n-2}s + \cdots + ts^{n-2} + s^{n-1}\right)$$
If also $0 < s < t$, then we see that
(16.2) $$t^n - s^n < (t-s)n t^{n-1}$$
which should suggest a calculus formula.

Our proof will use the familiar bisection algorithm. We need positive rational numbers $a < b$ such that $a^n \leq r \leq b^n$. If $r \geq 1$, we can use $a = 1/2$ and $b = r$. If $r < 1$, use $a = r$ and $b = 1$. Let $\delta = b - a$.

Now iterate the following: let $c = (a+b)/2$. If $c^n \le r$, replace a by c; otherwise, replace b by c.

At each step of the foregoing we have $a^n \le r \le b^n$, and at the k-th iteration, $0 < b - a < \frac{\delta}{2^k}$, as is easily verified by induction. Also,

$$0 \le b^n - r \le b^n - a^n$$

We will show that $b^n - r$ gets arbitrarily close to 0 by showing that $b^n - a^n$ gets arbitrarily close to 0. To do this, we use equation (16.2). Let B be the original value of b, so that $b \le B$ at each step. Then

$$b^n - a^n < (b-a)nb^{n-1} \le (b-a)nB^{n-1} = \frac{\delta}{2^k}nB^{n-1}$$

By Proposition 16.5 we can choose k so that the right side is as small as we wish. □

Proposition 16.6 hints that the rationals may be good enough for arbitrarily accurate numerical approximations, even though they do not contain exact solutions to some equations. The question whether approximations are "good enough" is extremely important. More food for thought.

When we constructed the rationals, we hinted at general questions concerning the nature of mathematical objects and their relation to each other. At the end of Chapter 1 we mentioned an article of Pourciau on questions of construction and foundation. The book of Dedekind mentioned in Chapter 2 is also relevant. Archimedes' property occurs in geometric form as Axiom 5 in his work *On the Sphere and the Cylinder*, translated into English in Heath's *The Works of Archimedes*, Dover 2002.

Problems

188. You are going to come up with a formal construction of the integers, starting only with the natural numbers. Thus, we are returning to the beginning of the course, before Chapter 2, and establishing some of the integer axioms in a new way. We will confine our construction to addition and ordering, leaving out multiplication.

Axioms. You may assume the following properties of the natural numbers: There is a set $\mathbb{N}$. (This set exists on its own; it is not the "positive integers," since there are no integers as yet!)

(a) There is $1 \in \mathbb{N}$.
(b) For $x, y \in \mathbb{N}$ there is $x + y \in \mathbb{N}$, and $x + y = y + x$. If $x, y, z \in \mathbb{N}$, then $(x + y) + z = x + (y + z)$.
(c) If $x, y, z \in \mathbb{N}$ and $x + y = z + y$, then $x = z$. (Note: we are *not* assuming that $-y$ exists, and there is no *subtraction*.)
(d) If $x, y \in \mathbb{N}$ then exactly one of the following is true: $x = y$ or $x < y$ or $y < x$. If $x, y, z \in \mathbb{N}$ and $x < y$, then $x + z < y + z$.

Idea of the construction. Informally: represent each integer as $a-b$, where $a, b \in \mathbb{N}$. Formally: for $a, b \in \mathbb{N}$, we represent $a - b$ by (a, b). (Note that you cannot speak of $a - b$ formally, since we don't have subtraction.)

Tasks in the construction. Do the following.

(a) Give a formal definition of $(a, b) \equiv (c, d)$ so that, informally, $(a, b) \equiv (c, d)$ tells us that $a - b = c - d$. Show, via your formal definition, that $\equiv$ is an equivalence relation.
(b) Give a formal definition of $\mathbb{Z}$.
(c) Define addition on $\mathbb{Z}$, and prove that it is commutative and associative, identify an identity element, and show that each integer has an additive inverse.

(d) Give a formal definition of $x < y$ for $x, y \in \mathbb{Z}$. Show that if $x, y \in \mathbb{Z}$ then $x = y$ or $x < y$ or $y < x$. Show that if $x, y, z \in \mathbb{Z}$ and $x < y$, then $x + z < y + z$.

(e) Find a copy of $\mathbb{N}$ inside $\mathbb{Z}$ and show that the copy has the same properties as $\mathbb{N}$ does (i.e. verify the axioms (a), (b), (c), and (d), given above).

Index

$A \iff B$, A if and only if B, 4
$A \cap B$, the intersection of A and B, 7
$A \cup B$, the union of A and B, 36
$A \Rightarrow B$, if A, then B, 3
$A \times B$, the Cartesian product: ordered pairs from A, B, 39
$C_G(x)$, centralizer of x in G, 91
D_6, a symmetry group, 61
D_8, a symmetry group, 61
E_A, identity function on A, 33
G/H, the cosets of H in G, 99
$G \cong K$, groups G, K are isomorphic, 125
$H \triangleleft G$, H is a normal subgroup of G, 107
K_4, Klein four-group, 65
Q_8, quaternion group, 75
R_2, the integers adjoin $\sqrt{2}$, 147
U_n, units group mod n, 63
$X \setminus Y$, the elements of X not in Y, 65
$Z(G)$, center of group G, 91
$[abc]$, a cycle, 46
$\mathbb{A}_P$, alternating group, 54

$\mathbb{N}$, the natural numbers (positive integers), 13
$\mathbb{S}_P$, the set of permutations of P, 45
$\mathbb{S}_n$, permutations of $1, \ldots, n$, 45
$\mathbb{Z}$, the integers, 11
$\mathbb{Z}_n$, the integers mod n, 25
$\ker(f)$, kernel of f, 120, 149
$\mathbb{Q}$, the rational numbers, 166
$\phi(n)$, totient function, 93
ϕ, the empty set, 6
$\subseteq$, is a subset of, 7
$\text{cl}(x)$, the conjugacy class of x, 135
$a \equiv b$, a is congruent to b, 25
$a \in B$, a is an element of the set B, 5
$a \notin B$, a is not an element of the set B, 5
$f : A \to B$, f is a function from A to B, 31
f^{-1}, inverse of f, 35
$o(x)$, order of group element x, 71
xH, a left coset of H, 98
x^{-1}, inverse of group element x, 58

abelian, 64
alternating group, 54

175

and (logic), 3
associative law for functions, 32
Aut(G), the set of automorphisms of G, 127
automorphism (of a group), 127
axiom, 1
Axiom of Choice, 41

bijection, 41
bijective, 41
binomial coefficients, 44
bounded above, 19
bounded below, 19

Cancellation Rule, 21
canonical homomorphism, 120
Cartesian product, 39
Cauchy's Theorem, 139
Cauchy's Theorem for Abelian Groups, 111
center (of a group), 91
centralizer of x (in the group G), 91
Chinese Remainder Theorem, 155
class equation, 138
closed, a subset under an operation, 90
commutative ring, 145
composite function, 32
conclusion (of implication), 3
congruent, 25
conjugacy class, 135
contrapositive, 4
converse, 4
coordinates, of an ordered pair, 39
Correspondence Theorem, 111
coset, 97

coset homomorphism, 122
cycle, 46
cyclic group, 69

dihedral group, 63
direct product of groups, 64
direct product of rings, 152
disjoint cycles, 47
disjoint sets, 36
divides, 23
Division Theorem, 23
domain, 145
domain (of a function), 31

edge, in a graph, 60
empty set, 6
equality of functions, 31
equality of ordered pairs, 39
equality of sets, 10
equivalence relation, 25
equivalent (statements), 4
even integer, 24
even permutation, 54
existential quantifier, 6

Fermat's Little Theorem, 158
field, 148
finite (set), 36
fixed points, 46
function, 31
function composition, 32
Fundamental Theorem of Abelian Groups, 131
Fundamental Theorem of Arithmetic, 83

GCD, 77
GCD Theorem, 77

generator (of a group), 69
graph of a function, 41
graph, finite undirected, 60
greatest common divisor, 77
group homomorphism, 119
group isomorphism, 124

hypothesis (of implication), 3

ideal, of a commutative ring, 150
identity element, 57
identity function, 33
image (of a function), 32
implication, 3
induction, 15
infinite order (of a group element), 73
infinite set, 43
injection, 41
injective, 41
integers, 11
intersection, 7
inverse (function), 35
inverse (of group element), 57
involution, 74
isomorphic (groups), 124

kernel of a homomorphism, 120, 150
Klein Four-Group, 65

Lagrange's Theorem, 99
least common multiple, 72
left coset, 98
lower bound, 19

maximal element, 19
maximal ideal, 154

minimal counterexample, 109
minimum, 14
mod, 25
modulo, 25
modulus, 25

N/C Theorem, 134
negation, 2
normal subgroup, 107
normalizer, of a subgroup in a group, 117

odd integer, 24
one to one (function), 33
onto (function), 33
operation, 40
or, logic, 3
order (of a group element), 71
order, of a finite set, 36
ordered pair, 39

parity (of a permutation), 53
Parity Theorem on permutations, 53
Peano's Axioms, 20
permutation, 45
pigeon-hole principle, 38
points, 45
preserves edges, 60
prime (natural number), 81
proper subgroup, 89
Pythagorean triple, 87

quantifiers, 4
quaternion group, 75
quotient, 23
quotient group, 108

range (function terminology deprecated), 32
rational numbers, 166
reflection, 62
remainder, 23
ring (commutative), 145
ring homomorphism, 149
rotation, 61

semi-direct product, 130
simple group, 114
square-free (natural number), 87
subgroup, 89
subset, 7
surjection, 41
surjective, 41
Sylow subgroup, 139
Sylow's Theorem (existence), 138
symmetry (of a graph), 60

totient function, 93, 155

union of sets, 36
units (of a ring), 148
units group mod n, 63
upper bound, 19

vacuous (implication), 3
vertex, in a graph, 60

well-ordering in the integers, 20
well-ordering of the natural numbers, 14

xor, exclusive or, 9

www.ingramcontent.com/pod-product-compliance
Lightning Source LLC
Chambersburg PA
CBHW081724170526
45167CB00009B/3688